Hiding the Decline

HIDING THE DECLINE

A.W. Montford

ISBN 978-1475293364

Set in Charter.

Cover design by the author. The chart shows the twentieth century divergence between some tree rings and the instrumental temperatures.

Printed and bound by Createspace

For Lesley

CONTENTS

LIST OF FIGURES

LIST OF TABLES

PREFACE

The events at the University of East Anglia in November 2009 had a profound and transformative effect on my life. Coinciding with the publication of *The Hockey Stick Illusion*, Climategate catapulted me into the front line of the Hockey Stick wars and brought me a measure of fame (or notoriety depending on one's view) if not of fortune. The success of *The Hockey Stick Illusion* was always going to be a hard act to follow and by deciding to set out the story of Climategate I have made my life doubly difficult. Many of those criticised in this book have set out to cover their tracks with a web of deceit and obfuscation and in this they have been moderately successful. The result of these efforts has been that the tale has been extraordinarily difficult to tell. I hope my efforts mean that it is somewhat easier to follow.

I have set out to write this record of what has happened in the last two years more out of a sense of public duty than because I think that I have a bestseller on my hands. Nevertheless, the story is an important one and, I think, one that carries many insights into the way the public sector functions.

The roots of the Climategate story are intertwined with that of the Hockey Stick graph, a tale that I have therefore been forced to retell in potted form in the first chapter of this new book. This has given me a chance to focus on those details that appeared insignificant at the time but now take on a new importance. I would advise those who are familiar with my last book not to skip over the new chapter so that the all the threads of the tale can be grasped.

Uncovering these complications has inevitably involved much detective work, a task that would have been far too great for anyone to complete on their own. Apart from myself, that task has fallen chiefly to Steve McIntyre and David Holland, to whom I am grateful for tips and clues and many useful discussions on the interpretation of the evidence.

Many others have helped out along the way, either with the text of the book or providing support along the way. Names that spring to mind Ross McKitrick, Josh Gifford, Peter Gill, Chris Horner, Doug Keenan, Richard Thomas, Jonathan Jones, Matt Ridley, Benny Peiser, Tony Newbery, David

Henderson, Anthony Watts, Steven Mosher, Tom Fuller and Don Keiller, as well as several others who have to remain anonymous. Thanks also to all those commenters at my blog who have suggested lines of inquiry, and also the many anonymous small donors who help support my blogging work and hence the investigations that have underpinned this book. If I have missed anyone out from this list, I apologise.

As with *The Hockey Stick Illusion*, Dr Lesley Montford and Dr Angela Montford both helped bring the book to fruition in their separate ways. To them, many thanks are due. Smaller members of the family kept me from my desk as much as possible, which was probably just as important in bringing the project to a successful conclusion.

AWM
Kinrossshire, 2012.

NOTES ON USAGES

As it is one of the most important allegations arising from the Climategate emails, I return several times to Phil Jones' infamous truncation of one of Keith Briffa's tree ring series. Readers will often see this incident referred as 'Mike's Nature trick', but I have adopted the usage of 'the trick to hide the decline'. In the often absurdly heated atmosphere of the global warming debate, criticism is sometimes levelled at those who place the latter expression in quotation marks. It is said to be unreasonable to imply that this is a direct quotation since in Jones' original email the words 'trick' and 'hide the decline' are separated by a parenthetical clause. To avoid this kind of unpleasantness I will use the expression without quotation marks – the trick to hide the decline – or 'the Trick' for short.

UK freedom of information laws are set out in two main pieces of legislation: the Freedom of Information Act 2000 and the Environment Information Regulations 2000. I will refer to 'FOI legislation' when I mean to refer to both of these at the same time.

We still have no idea of the identity of the person who disclosed the Climategate emails (nor indeed if they were a hacker or a whistleblower). I refer to the person involved as 'he', but this should not be taken as any indication that the field of suspects has been narrowed in any way.

Finally, as in *The Hockey Stick Illusion*, this book includes many extracts from the email correspondence of the main characters. In general I have chosen to correct spelling and punctuation errors in these in order to avoid having to excuse myself of each one.

CHAPTER 1

THE HOCKEY STICK

The roots of this story are long and tangled and sometimes hard to grasp, but if one were to point to an element that is key to understanding the events at the University of East Anglia at the end of 2009, it is the story of the Hockey Stick graph. I related this extraordinary tale in my last book, *The Hockey Stick Illusion*, but it is nevertheless important to revisit at least some of the details here, to enable new readers to understand the background to the story and also to focus on those elements that turned out to be important in what followed.

The Hockey Stick is a graph of global temperatures for the last millennium, reconstructed from tree rings and other so-called proxy data, its name coming from its remarkable shape, a long flat 'handle' representing comparatively stable temperatures in earlier centuries, followed by a dramatic uptick – the 'blade' – representing the effect of industrialisation on temperatures in the twentieth century. It was, for a time, the most important graph in the world, its message of unprecedented warmth at the end of the twentieth century a vital part of the campaign to persuade a doubting public that mankind had changed the world's climate. The graph was the work of a young American climatologist named Michael Mann, and it was controversial from the moment it appeared in the journal *Nature* in 1998, with its publication prompting a fierce debate over whether Mann and his co-authors, Raymond Bradley and Malcolm Hughes, had spliced two different datasets or had merely 'overlaid' them.

Within days of its publication the Hockey Stick graph was being extensively cited and soon afterwards it was picked up by the Intergovernmental Panel on Climate Change (IPCC), the body that issues periodic assessments of the world's climate. The IPCC viewed Mann's findings as highly significant and gave the Hockey Stick top billing in its Third Assessment Report, showing it no less than six times across the various parts of the report. Despite, or perhaps because of, the importance attached to the graph, it quickly became a target for sceptics, who were suspicious of Mann's findings and the relentless way in which the graph was being promoted as evidence of manmade global warming.

SOON AND BALIUNAS

One of the first attempts to rebut the Hockey Stick paper was made by two Harvard astrophysicists, Willie Soon and Sallie Baliunas, who published a paper in 2003 in the journal *Climate Research*, concluding that Mann had been incorrect and that temperatures at the end of the twentieth century had been well within the normal range of climate variability.[1] In fact, they said, modern temperatures were likely to have been surpassed during the Medieval Warm Period. Soon and Baliunas's paper was therefore a direct challenge to both the Hockey Stick itself and to the IPCC report, which placed great emphasis upon its dramatic message of unprecedented warmth in the twentieth century. It was clear from the moment the Soon and Baliunas paper appeared that there would be a response.

One of the first mainstream climatologists to notice the new paper was Tim Osborn, a scientist at the Climatic Research Unit (CRU) at the University of East Anglia (UEA) in the UK. Realising that the paper had the potential to strike a serious blow against the 'consensus' position that modern temperatures were unprecedented, Osborn quickly brought it to the attention of the unit's director, Professor Phil Jones. Shortly afterwards, Jones forwarded the news to a group of senior colleagues. Although it is not clear exactly who received Jones' message, the recipients of later emails in the thread included Michael Mann as well as several other members of the tightly knit group of climatologists who would later become known as the 'Hockey Team': Jonathan Overpeck, a senior IPCC author, and Keith Briffa, a climatologist who worked alongside Jones at CRU. The message was as follows:

> Tim Osborn has just come across this [paper]. Best to ignore probably, so don't let it spoil your day. I've not looked at it yet. It results from this journal having a number of editors. The responsible one for this is [Chris de Freitas,] a well-known skeptic in [New Zealand]. He has let a few papers through by [sceptics] in the past.[2]

This first message has something of a tone of weary resignation: as Jones explained, he had already complained several times to *Climate Research* about its willingness to publish sceptic papers, but the editor, Hans von Storch, had been unimpressed. However, the Soon and Baliunas paper was clearly much more serious in effect than any of these earlier episodes, and Jones and his colleagues started to discuss taking much firmer steps to bring the journal to heel – steps that may have crossed an ethical line.

Jones' next message started out in innocent enough fashion, with a detailed description of his concerns over the new paper. In particular, he felt that Soon and Baliunas should have discussed the question of whether the Medieval Warm Period happened at the same time in different places around the world. At first sight, this would appear to be a fairly run-of-the-mill difference of scientific opinion – one that would, in the normal state of affairs, lead to the submission of a critical comment to the journal. However, as it continued, Jones' email started to take on an uglier tone and he told his colleagues that the paper was 'appalling':

> Writing this I am becoming more convinced we should do something – even if this is just to state once and for all what we mean by the [Little Ice Age and Medieval Warm Period]. I think the skeptics will use this paper to their own ends and it will set [paleoclimatology] back a number of years if it goes unchallenged. I will be emailing the journal to tell them I'm having nothing more to do with it until they rid themselves of this troublesome editor. A CRU person is on the editorial board, but papers get dealt with by the editor assigned by Hans von Storch.[2]

Remarkably then, Jones appears to have been suggesting a boycott of the journal if it published sceptic papers. Mann was in complete agreement and he expanded on Jones' hint that there might be something untoward going on at the journal:

> The Soon and Baliunas paper couldn't have cleared a 'legitimate' peer review process anywhere. That leaves only one possibility – that the peer-review process at *Climate Research* has been hijacked by a few skeptics on the editorial board.[2]

In fact, Mann was even more suspicious than Jones, pointing the finger of doubt at von Storch as well as de Freitas:

> My guess is that von Storch is actually with them (frankly, he's an odd individual, and I'm not sure he isn't himself somewhat of a skeptic himself)...[2]

As Mann went on to suggest, one of the main criticisms of the sceptics was that they did not publish their findings in the peer-reviewed journals, and he was determined that something should be done to stop Soon and Baliunas winning a public relations coup by actually doing so.

This was the danger of always criticising the skeptics for not publishing in the 'peer-reviewed literature'... they found a solution to that – take over a journal! So what do we do about this? I think we have to stop considering *Climate Research* as a legitimate peer-reviewed journal. Perhaps we should encourage our colleagues in the climate research community to no longer submit to, or cite papers in, this journal. We would also need to consider what we tell or request of our more reasonable colleagues who currently sit on the editorial board...[2]

For a time the Hockey Team discussed the possibility of a more orthodox response – a published rebuttal in another journal. However, some members of the team were still keen on something more radical. Over the following weeks the conversation expanded to include Mike Hulme, the director of the Tyndall Centre, the UK's national centre for climate research, which was also sited on the UEA campus. Also involved were a number of scientists based in Australia and New Zealand. Hulme in particular was keen to make the response decisive – his idea was to prompt a mass resignation of the editors of *Climate Research*, and he noted that Jones had carried through on his threat and was already refusing to peer review any papers for the journal. Jim Salinger, a scientist at New Zealand's National Institute of Water and Atmospheric Research, was also one of the hawks. In his emails he was alluding darkly to de Freitas's right-wing views and he offered to compile a dossier of his journalism as evidence. Jones said that he intended to put even more pressure on von Storch the following week:

... I'll be telling him in person what a disservice he's doing to the science and the status of *Climate Research*. I've already told Hans I want nothing more to do with the journal.[3]

Mann was grateful for his colleagues' support of his paper and said that he was encouraged at the prospect of some sort of action being taken, suggesting that a complete boycott of the journal would be in order.

I believe that a boycott against publishing, reviewing for, or even citing articles from *Climate Research* is certainly warranted.[3]

Shortly afterwards, Hulme sent a letter to a group of *Climate Research* editors, explaining the strength of feeling among mainstream paleoclimatologists about Soon and Baliunas's paper, which he said was 'just crap science that should never [have] passed peer review'.[3] He explained that communications managers at the Tyndall Centre had backed his idea of a mass resignation.

While Hulme was pursuing this project, other members of the Hockey Team were considering other possibilities. Barrie Pittock, a scientist at Australia's national research institute, CSIRO, circulated a long list of ideas, including branding dissenting editors as 'rogue editors', taking legal action – he noted that it might be possible to obtain financial support for this – and returning to the idea of a boycott to bring recalcitrant journals back into line.[4]

A few days later, Salinger wrote a long email to the rest of the Team, copying it to a larger group of colleagues in Australia as well as Rajendra Pachauri, the head of the IPCC. In it, he explained the still more dramatic steps he had in mind:

> I have had thoughts also on a further course of action. The present Vice Chancellor of the University of Auckland, Professor John Hood (comes from an engineering background) is very concerned that Auckland should be seen as New Zealand's premier research university... My suggestion is that a band of you review editors write directly to Professor Hood with your concerns. In it you should point out that you are all globally recognized top climate scientists. It is best that such a letter come from outside NZ and is signed by more than one person.[5]

Salinger helpfully provided a suggested text,

> We write to you as the editorial board (review editors??) of the leading international journal *Climate Research* for climate scientists... We are very concerned at the poor standards and personal biases shown by a member of your staff...
>
> When we originally appointed... to the editorial board we were under the impression that they would carry out their duties in an objective manner as is expected of scientists world wide. We were also given to understand that this person has been honoured with science communicator of the year award, several times by your... organisation.
>
> Instead we have discovered that this person has been using his position to promote 'fringe' views of various groups with which they are associated around the world. It perhaps would have been less disturbing if the 'science' that was being passed through the system was sound. However, a recent incident has alerted us to the fact

that poorly constructed and uncritical work has been allowed to enter the pages of the journal. A recent example has caused outrage amongst leading climate scientists around the world and has resulted in the journal dismissing (??)...from the editorial board.

We bring this to your attention since we consider it brings the name of your university and New Zealand into some disrepute. We leave it to your discretion what use you make of this information.

The journal itself cannot be considered completely blameless in this situation and we clearly need to tighten some of our editorial processes; however, up until now we have relied on the honour and professionalism of our editors. Sadly this incident has damaged our faith in some of our fellow scientists. Regrettably it will reflect on your institution as this person is a relatively senior staff member.[5]

Tom Wigley, a former head of CRU, appeared somewhat uncomfortable with what was being discussed. However, while he said that he was convinced that de Freitas was selecting reviewers sympathetic to sceptic views, he maintained that 'a barrage of ad hominem attacks or letters' was not the way forward,[6] and in fact his objections may have carried the day since there is no evidence that the Hockey Team ever acted on Salinger's more extreme ideas. Wigley also felt that some of the blame should be apportioned to von Storch:

Hans von Storch is partly to blame – he encourages the publication of crap science 'in order to stimulate debate'. One approach is to go direct to the publishers and point out the fact that their journal is perceived as being a medium for disseminating misinformation under the guise of refereed work. I use the word 'perceived' here, since whether it is true or not is not what the publishers care about – it is how the journal is seen by the community that counts.[7]

Disturbingly, Wigley's message carries a hint that he wanted *Climate Research* closed to the sceptics whether the misdemeanours the Hockey Team saw in it were real or not, and he went on to reveal that von Storch was also in his sights:

[Mike Hulme's] idea to get editorial board members to resign will probably not work – must get rid of von Storch too, otherwise holes will eventually fill up with [sceptics].[7]

Mann was particularly keen on this idea, emphasising the political importance of taking action against *Climate Research*.

> This latest assault uses a compromised peer-review process as a vehicle for launching a scientific disinformation campaign (often vicious and ad hominem) under the guise of apparently legitimately reviewed science...Much like a server which has been compromised as a launching point for computer viruses, I fear that *Climate Research* has become a hopelessly compromised vehicle in the skeptics' (can we find a better word?) disinformation campaign, and some of the discussion that I've seen (e.g. a potential threat of mass resignation among the legitimate members of the...editorial board) seems, in my opinion, to have some potential merit.

> This should be justified not on the basis of the publication of science we may not like of course, but based on the evidence ... that a legitimate peer-review process has not been followed by at least one particular editor.[8]

Interestingly, Mann also revealed a belief that the scientific literature was largely closed to sceptic views, with only a handful of journals still permitting dissenting views to be aired:

> While it was easy to make sure that the worst papers...didn't see the light of the day at [*Journal of Climate*], it was inevitable that such papers might slip through the cracks at e.g. [*Geophysical Research Letters* (GRL)] – there is probably little that can be done here, other than making sure that some qualified and responsible climate scientists step up to the plate and take on editorial positions at GRL.[8]

Shortly afterwards, Jones reported to the Team on his meeting with von Storch. Interestingly he wrote two separate emails – one to Hulme and one to Mann. To his UEA colleague he explained that von Storch was now onside and would be writing to the publisher to help the team identify the scientists de Freitas had chosen to peer review the Soon and Baliunas paper.[9] The same day, he revealed rather more to Mann, telling him that von Storch had actually gone much further and had indicated that de Freitas would be forced out of his position.[10] The news was soon spreading. Clare Goodess, one of Jones' colleagues at CRU and a member of the *Climate Research* editorial board, apparently told Mann that de Freitas would soon be gone.[11] This appears to represent an extraordinary breach of privacy, with de Freitas's position at *Climate Research* being freely discussed by a variety of scientists, at least some of whom were unconnected with the journal.

What happened next is not entirely clear, but a partial record of events has been given by Goodess who, although she had been party to much of the correspondence among the members of the Hockey Team, made no mention of this in her account of events. According to her account, she and von Storch were sent 'numerous unsolicited complaints and critiques of the paper from many leading members of the international [paleoclimate] and historical climatology community'.[12] Another complaint, which Goodess did not mention, was sent direct to the journal's publisher, Otto Kinne, by Hulme.[13] Although the contents of this message have not been made public, much of its content can be determined from de Freitas's defence of the accusations Hulme had made about him – that he was politically motivated and that he was giving an easy ride to sceptic papers.[13] In response, de Freitas argued that he had invited no fewer than five scientists to review the paper, although one had been too busy to take part. However, he had not selected the five names himself, but instead had passed this duty on to a scientist with expertise in the field. As Kinne noted in a subsequent letter to his editorial team this all appeared to be entirely normal – in fact he personally reviewed de Freitas's files and could find nothing untoward – de Freitas had been targeted by critics in the past and was very careful to document his work thoroughly so as to be able to defend his decisions.[14]

Mann was unconvinced and, in what was later to emerge as a pattern in his behaviour, set about raising the temperature of the debate:

> It seems clear we have to go above [Kinne]. I think that the community should, as Mike [Hulme] has previously suggested in this eventuality, terminate its involvement with this journal at all levels – reviewing, editing, and submitting, and leave it to wither way into oblivion and disrepute.[15]

Wigley was in complete agreement with Mann's provocative plan, and started to plan ahead, asking...

> ...what would be our legal position if we were to openly and extensively tell people to avoid the journal?[15]

Jones' contribution to the conversation also adds some important details about the nature of de Freitas's review. Firstly Jones said that he thought that the paleoclimatologist who had selected the reviewers on de Freitas's behalf was Anthony Fowler – a researcher from New Zealand.

Fowler is not known to be a sceptic, so Jones' news should have provided considerable reassurance to the Hockey Team members. But if this information was insufficient to convince them, they should certainly have had their concerns eased by the news that the scientist who had been too busy to get involved in the review was none other than Ray Bradley, Mann's co-author on the Hockey Stick papers – hardly someone who would be chosen by an editor looking to 'fix' the result of the peer review in favour of the sceptics. Remarkably, however, these revelations did nothing to change the views of the Hockey Team, and Jones even suggested that a negative review from Bradley would have made no difference to de Freitas' decision to publish. The reasons why Mann and Jones and their colleagues were so keen to proceed with action against the journal are unclear but, as we will see, there were hints of another, non-scientific agenda operating in the background.

At the same time as they were emailing Kinne and de Freitas, the Hockey Team scientists were busy preparing a more legitimate response to the Soon and Baliunas paper – a formal rebuttal, which they intended to submit to the journal *Eos*, where the editorial staff would be sympathetic to their position. The team of authors assembled for the rebuttal paper was a *Who's Who* of the Hockey Team: Mann and his co-authors Bradley and Hughes, Jones, Briffa and Osborn from CRU, and several others whose names will become familiar over the course of this story: Kevin Trenberth, Jonathan Overpeck and Caspar Ammann. Lastly, there was Michael Oppenheimer, a Princeton professor and long-time advisor to a green advocacy group, the Environmental Defense Fund. Towards the end of July Mann and Oppenheimer wrote to the rest of the authors of the *Eos* piece suggesting that they should also write a joint letter to the US Senate because, as Mann put it, there was a 'continued assault on the science of climate change by some on Capitol Hill'.[16] As the email makes clear, Soon and Baliunas's work was having an impact on the political process in the USA, and Mann was determined to prevent this happening. In other words, it may not have mattered who had reviewed the Soon and Baliunas paper; it was necessary for it to be discredited.

Wigley agreed that politicians were making use of Soon and Baliunas' work and wondered if there was a possibility that the American Geophysical Union (AGU) and the American Meteorological Society might be recruited to help in discrediting it. Alternatively he wondered if Soon and Baliunas's university – Harvard – could be persuaded to distance themselves from the two sceptics in some way. Mann concurred and noted that the editor of *Eos* might be prevailed upon to get the AGU to make a statement. He said that Oppenheimer had also suggested that they might also be able to get the editor of *Science*, Donald Kennedy, to have the American Association for the Advancement of Science make a stand against the paper. However, most of the Hockey Team members were lukewarm about the idea of writing to the Senate, and a decision was taken to concentrate on the *Eos* article instead.

Meanwhile, in an attempt to put affairs at *Climate Research* back on an even keel, Kinne decided to promote Hans von Storch to the position of editor-in-chief, with a brief to oversee the review process and ensure its integrity. However, this idea quickly backfired. Just days before Soon and Baliunas were due to speak about their paper to legislators in the Senate, von Storch circulated the draft of an editorial that he said he would publish in the next issue of the journal.[17] He proposed telling readers that the Soon and Baliunas paper was flawed and that the peer review process at the journal had been inadequate to stop it from appearing.

The editorial board, however, were by no means united behind this idea and at least one of its members objected strongly, as he explained in an email to von Storch:

> A paper has been published that some people disagree with ... the authors have responded. Isn't this the nature of the same scientific process that has worked just fine for centuries? Many papers have been published with which I have disagreed, but I never viewed the 'process' to be flawed. Honest scientists have differences of opinion. That is clearly the case here. You should know that I know the parties on BOTH sides of this particular issue and am not taking sides.
>
> I cannot agree with your editorial since, in my view, there is no problem with the peer-review process.[17]

The result was inevitable. Shortly afterwards, and neatly coinciding with Soon and Baliunas's appearance on Capitol Hill, von Storch and several members of the editorial board at *Climate Research* resigned, no doubt to the horror of the politicians who had invited the two sceptics to speak, and to delight of their opponents.

Soon and Baliunas's critique of the Hockey Stick lapsed into obscurity, but in fact the Hockey Team was not quite finished with the two sceptics. A few months later, despite all the effort expended on discrediting them, an issue of the journal *Progress in Physical Geography* featured a review article on global warming written by Soon and Baliunas. This appearance, so soon after their humiliation on Capitol Hill appears to have been too much for Hulme, who wrote to the journal's editor in no uncertain terms:

> I am writing to resign from my position as Editorial Adviser for the journal Progress in Physical Geography...I reached this decision after seeing the September 2003 issue of the journal in which I noticed that Willie Soon and Sallie Baliunas have been asked to provide the annual progress reports for 'global warming' for the journal and after reading their first contribution.[18]

And with that, the work of the Hockey Team appeared to be complete, as least as far as Soon and Baliunas were concerned.

MCINTYRE

Shortly after the Soon and Baliunas affair had ended in disappointment for the sceptic community, a new figure began to look at the Hockey Stick paper. Steve McIntyre was a semi-retired mining consultant from Toronto who had begun to investigate Mann's paper on a whim. Armed with formidable mathematical skills and a dogged determination, he quickly unearthed a raft of problems with Mann's data, findings which he published later in 2003 in a paper co-authored with his fellow Canadian, economist Ross McKitrick. The paper, which we will refer to as MM03, appeared in an obscure journal called *Energy and Environment*, which had a reputation as the sceptics' journal of choice, since its editorial board were sympathetic to dissident voices.

On the eve of the publication of MM03, Mann circulated an email to several of his colleagues, attaching a commentary about McIntyre and McKitrick's paper. This remarkable document, written by an unnamed author, shows that despite their outwardly aggressive stance towards anyone who questioned mainstream climate science, some of Mann's colleagues seem to have been privately impressed by MM03. Indeed the commentary suggested that their main conclusion – that the Hockey Stick graph's story of little medieval warmth was changed by minor corrections to the underlying database – was already well known to 'those who understand Mann's methodology'.[19] However, the author of the message was concerned that Mann's hot temper was going to get him into trouble once he saw the paper, and these concerns turned out to be well founded. Despite being aware that at least one of his associates agreed with McIntyre and McKitrick, Mann's immediate reaction was not to challenge MM03 through a response in the scientific literature, but instead to question the integrity of *Energy and Environment*. He suggested that the journal was 'a shill for industry', and on this basis dismissed the paper in its entirety rather than engaging with its criticisms of his work. As Mann told his colleagues:

> The important thing is to deny that this has any intellectual credibility whatsoever and, if contacted by any media, to dismiss this for the stunt that it is.[19]

With the assistance of a sympathetic journalist, Mann proceeded to issue a strongly worded denunciation of McIntyre and McKitrick. A decision was also taken to issue a further informal rebuttal via CRU, since the UK wing of the Hockey Team could then be presented to the public as neutral arbiters in the dispute.[19] In due course a response was posted on Tim Osborn's website at CRU.[20]

The first skirmishes between Mann and the two Canadians were inconclusive, but Mann's hot-headed response to the publication of MM03 had led McIntyre to another extraordinary series of discoveries about the way the Hockey Stick paper had been put together. Mann's database included a number of hockey-stick-shaped tree-ring series derived from bristlecone pine trees, despite this species being widely recognised in the literature as being contaminated with a non-climatic signal. Then, to make it worse, it had been discovered that part of Mann's computer algorithm would pick hockey-stick shaped series out of the database and overweight them in the final result – in other words, even if there were only a few hockey-stick-shaped series in the database, the final reconstruction would look like a hockey stick. Together, McIntyre's damning findings looked as though they would break Mann's creation once and for all.

One of the main criticisms of MM03 had been the fact of its publication in *Energy and Environment*, which Hockey Team supporters claimed was not even a science journal – in fact, some had suggested that MM03 had not actually been peer reviewed at all.* With this in mind, it had been important for McIntyre and McKitrick to get their new findings published in a more mainstream journal and they had initially submitted

---

*Note, however, the email in which Osborn says that *Energy and Environment* had disputed this and that one scientist – Fred Singer – has identified himself as having reviewed the paper.[21]

their manuscript to *Nature,* the journal that had published the Hockey Stick paper in the first place. However, after a protracted review process, the journal issued a rejection, apparently on the somewhat surprising grounds of lack of space.

Despite this, just as with MM03, some of Mann's closest colleagues appear to have been favourably impressed by the case put forward by the two sceptics. Wigley told Jones that McIntyre's findings appeared valid and that Mann's was 'a very sloppy piece of work'.[22]* Jones however, was deep in the Mannian mindset by this time, and told Wigley that everything that McIntyre and McKitrick were saying was wrong, 'a complete distortion of the facts'.[23]

THE SAIERS AFFAIR

It must have been clear to Mann and his colleagues that McIntyre and McKitrick would not be put off by this setback and that they would try to publish their findings elsewhere. Soon afterwards their fears were proved correct when they learned that the Canadians' paper – which we will refer to as MM05 – had been accepted by GRL, one of the journals where the Hockey Team thought that that sceptic papers might 'slip through the cracks'.

Mann was furious, but he was not someone to give up easily and he made a desperate attempt to put a spanner in the works. Towards the end of January 2005 he telephoned Steve Mackwell, the editor-in-chief of GRL, complaining that he had not been allowed to comment on the paper and apparently trying to delay its publication. Although we cannot know precisely what Mann said, we can surmise many of these details from Mackwell's reply:

> While I do agree that this manuscript does challenge (somewhat aggressively) some of your past work, I do not feel that it takes a particularly harsh tone. On the other hand, I can understand your reaction. As this manuscript was not written as a comment, but rather as a full-up scientific manuscript, you would not in general be asked to look it over. And I am satisfied by the credentials of the reviewers. Thus, I do not feel that we have sufficient reason to interfere in the timely publication of this work.

---

*It is not entirely clear whether Wigley was discussing McIntyre's submission to *Nature* rather than MM03, although the date of the email makes the former the more likely. If so, then Mann's having circulated the manuscript to third parties would probably represent a breach of peer-review confidentiality.

However, you are perfectly in your rights to write a comment, in which you challenge the authors' arguments and assertions.[24]

Mann's reaction to this gentle rebuff was not to do as Mackwell had suggested and argue his case in a comment submitted for publication, but once again to assume that sceptics had 'captured' the journal in some way. However, his attention was drawn not to Steve Mackwell, but to his deputy, James Saiers, who had been the editor responsible for handling the McIntyre and McKitrick paper. Saiers had formerly worked at the University of Virginia alongside a noted sceptic climatologist, Pat Michaels, and Mann found this association to be highly suspicious, as he explained in an email to several of his colleagues:

> Just a heads up. Apparently, the contrarians now have an 'in' with GRL. This guy Saiers has a prior connection [with] the University of Virginia Dept. of Environmental Sciences that causes me some unease. I think we now know how [sceptic papers] have gotten published in GRL.[24]

Tom Wigley, the former head of CRU who had told Jones that McIntyre's new findings looked valid, was one of those who responded, but now presented a rather different take on the paper:

> This is truly awful. GRL has gone downhill rapidly in recent years. I think the decline began before Saiers...I have had some unhelpful dealings with him recently with regard to a paper...I got the impression that Saiers was trying to keep it from being published. Proving bad behavior here is very difficult. If you think that Saiers is in the greenhouse skeptics camp, then, if we can find documentary evidence of this, we could go through official [AGU] channels to get him ousted. Even this would be difficult.[24]

Mann appeared to agree with Wigley that Saiers should be dealt with in some way and suggested a possible course of action:

> It's one thing to lose *Climate Research*. We can't afford to lose GRL. I think it would be useful if people begin to record their experiences [with] both Saiers and potentially Mackwell (I don't know him – he would seem to be complicit [in] what is going on here).[24]

With Mackwell refusing to accept Mann's protests, there was little that the Hockey Team could do about the new paper. But they were not going to suffer this setback in silence and when a journalist from the *New York Times* contacted Mann for a comment, he was told that the paper was 'pure scientific fraud'.[25]

When MM05 was published at the start of 2005, there was something of a media storm, with the finding that the Hockey Stick was flawed making headlines around the world. McIntyre was even profiled on the front page of the *Wall Street Journal*. The media interest was driven not just by the scientific importance of the Hockey Stick but by its prominence in the IPCC reports and in the resulting political debates. So if McIntyre and McKitrick's new paper was a problem for Mann it was potentially even worse for the IPCC. Having relentlessly promoted the Hockey Stick in the Third Assessment Report back in 2001, it would now suffer a hugely embarrassing loss of face if it was forced to admit that Mann's paper had been wrong. To make things even more difficult, work on the Fourth Assessment Report was already under way, so if anything was to be done about MM05 before the report appeared it would have to be happen in fairly short order. It was not long before the Hockey Team swung into action again.

WAHL AND AMMANN

In May 2005, on the same day that McIntyre was due to make a rare public appearance at a Washington think tank, the source of the Hockey Team's response became clear. Just hours before McIntyre was due to speak, a press release was issued by the US National Center for Atmospheric Research on behalf of two of its scientists. One of these – Caspar Ammann – we have already met as one of the co-authors to the proposed rebuttal to Soon and Baliunas in *Eos*; the other, Eugene Wahl, was also a core member of the Hockey Team. The press release announced the submission of two articles for publication. The first was a paper for publication in the journal *Climatic Change* ('the paper'), which Wahl and Ammann claimed exactly replicated the Hockey Stick and demonstrated that Mann's findings were sound. The second was a comment on MM05 for publication in GRL ('the comment'), which they said would show that McIntyre and McKitrick's critique of the Hockey Stick was baseless. Perhaps there would be time to save the Hockey Stick for the Fourth Assessment Report after all.*

---

*Interestingly, there is some evidence that at the time MM05 was published, Wahl and Ammann were in the process of completing a paper rebutting MM03.[26] This earlier

However, while it was one thing for Wahl and Ammann to claim they had refuted McIntyre and McKitrick, it was quite another to actually do so in practice, for the simple reason that their two papers actually *supported* many of the Canadians' main criticisms of the Hockey Stick, in particular their claim that the Hockey Stick was statistically unreliable. At stake were the so-called verification statistics: numbers that give statisticians a feel for how much reliance they can place on a result. There were two main verification measures at issue, the R^2 and the RE. McIntyre and McKitrick had shown that the Hockey Stick failed the R^2, which is a routine measure used by statisticians. Mann, however, had been arguing for the use of the more obscure RE test, which was unknown outside the field of climatology, but McIntyre had shown that the Hockey Stick actually fell at this hurdle too. Since Wahl and Ammann claimed to have exactly replicated Mann's paper, this could only mean that their version of the Hockey Stick must have failed its verification statistics as well. In other words, both Mann's graph and Wahl and Ammann's replication of it were statistically unreliable. What was worse for the Hockey Team, Wahl and Ammann had published all their data and computer code online, so the details and proof that McIntyre and McKitrick were correct were there to be seen by anyone who took the trouble to look. So, no matter what Wahl and Ammann told the press they had done, they would not be able to hide the facts for long. Shortly afterwards, their tricks were to be exposed.

Since Wahl and Ammann were criticising their work, GRL invited McIntyre and McKitrick to provide a written response. This would be sent out to the peer reviewers alongside the comment, to allow them to assess both sides of the argument. When the invitation to respond arrived, the two Canadians held nothing back and issued a strongly worded denunciation of Wahl and Ammann's work, demonstrating a series of flaws in the science and some highly questionable ethics as well. In fact, so damning was the rebuttal that when he saw it, the GRL editor, James Saiers, decided that he could not credibly send Wahl and Ammann's comment out for peer review and took an editorial decision to reject it entirely.

The rejection must have been a crushing blow to Wahl and Ammann. Not only was their purported refutation of McIntyre and McKitrick in tatters, but their replication of the Hockey Stick – their paper in *Climatic*

---

rebuttal was presumably withdrawn before it could appear so that they could focus on the more serious findings in MM05.

Change – was potentially in ruins too because it relied for some of its key statistical arguments on the GRL comment. Even if they could get the paper through peer review in time, it would be simple for McIntyre to point out that it relied on the comment, which had already been rejected.

The Hockey Team's plan to save the graph for the IPCC looked as though it would come to nothing and, with the deadline for papers to be included in the Fourth Assessment Report looming just months away, there was almost no time left to do anything about it.

THE IPCC DEADLINE

The job of dealing with the Hockey Stick issue in the IPCC report was the responsibility of the coordinating lead authors on the paleoclimate chapter, Jonathan Overpeck and Eystein Jansen.* The two men had the unenviable task of coordinating an international team of authors and producing a chapter that covered climate history over many different timescales. And if this were not difficult enough, they also had to deal with the fallout from the most heated dispute of the whole report. The two years of the IPCC review promised to be interesting ones for the author team.

Because of the complexity of the chapter they were supervising, Overpeck and Jansen had divided the work up between the members of the team. The author who was given responsibility for the millennial temperature reconstructions – including the Hockey Stick – was, like Overpeck, a core member of the Hockey Team and a name we have come across already: the CRU's Keith Briffa.[27]

Briffa can have been under few illusions about the importance of the task he had been given, but if he harboured any doubts the email he received from Overpeck would have swept them aside. As Overpeck put it, Briffa's section was to be 'the big one' and he went on to explain that, exceptionally among the author team, Briffa would be allowed to expand beyond his initial word allocation if it helped him to produce a compelling section that could be included in the Summary for Policymakers – the executive summary of the report.[28] However, it was also clear that there would be no hiding from the Hockey Stick controversy. Even while the first draft – known as the Zero Order Draft – was being pre-

---

*We have already met Overpeck as one of the Hockey Team members who was involved in the discussion about how to react to the Soon and Baliunas paper.

pared, Briffa had been sent an email by David Rind, one of his fellow lead authors:

> [McIntyre and McKitrick] claim that when they used [the statistical procedure used by Mann, but with random data series], it always resulted in a 'hockey stick'. Is this true? If so, it constitutes a devastating criticism of the approach; if not, it should be refuted. While IPCC cannot be expected to respond to every criticism a priori, this one has gotten such publicity it would be foolhardy to avoid it.[29]

With the millennial temperature reconstructions intended to form the centrepiece of the paleoclimate chapter, Wahl and Ammann's attempt to save the Hockey Stick was clearly going be a matter of intense interest to the team at the IPCC. At the start of July Overpeck emailed Ammann to ask after the paper's progress, saying that it was 'most important' that it be in press by the end of the month – the deadline for inclusion in the First Order Draft.[30] There are no records of the reaction of any of the scientists to the rejection of Wahl and Ammann's comment on MM05 a few weeks earlier, but it is possible that none of them – Briffa and Overpeck included – were aware of how critical the arguments in the comment were to the case put forward in the paper. Wahl, however, must have realised just how difficult it was going to be to get both the paper and the comment through peer review at all, let alone before the IPCC report was issued. He wrote back to Overpeck, perhaps somewhat nervously, seeking clarification of exactly when the final deadline for submission of papers was to be.

Fortunately, an opportunity for Wahl to extricate himself from his predicament was soon to present itself. Shortly after his email exchange with Mann, Mackwell had come to the end of his term of office as editor-in-chief of GRL and was replaced by a new man – Professor Jay Famiglietti of the University of California. With the new regime came a new opportunity to save the Hockey Stick.

The first signs that something unusual was happening came when Famiglietti announced in a magazine interview that he had decided to take personal responsibility for MM05 and its responses, including Wahl and Ammann's comment, apparently on the grounds that the McIntyre and McKitrick paper had been so controversial. Saiers – the man the Hockey Team had discussed ousting just a few months earlier – was somewhat ignominiously to be pushed aside.* Shortly afterwards it emerged that Wahl and Ammann's comment had been resubmitted to GRL and that now, only a few days later, it was 'pending final acceptance'. This was extraordinary, since critical comments submitted to the journal were supposed to be sent out to reviewers accompanied by a response from the authors who were being criticised. In order to avoid another devastating critique from McIntyre and McKitrick, Famiglietti had been forced to break his journal's own rules, simply failing to allow the two Canadians to have their say. Having already accepted the comment, Famiglietti added insult to injury by belatedly inviting McIntyre to submit a response.

The integrity of the peer review process had been shattered in the scramble to save the Hockey Stick, but as Mann later commented to Jones and Osborn, the 'leak at GRL' had been 'plugged'.[31]

THE TIMETABLE

The coup at GRL and Famiglietti's acceptance of the comment that Saiers had rejected had brought all of Wahl and Ammann's statistical arguments back into play.[32] However, as we have seen, Wahl and Ammann's *paper* – the replication of the Hockey Stick – failed both of the main statistical tests used to assess its reliability and it was therefore unclear how the resurrection of the *comment* was going to help get it through its peer review. However, by accepting the comment without a response from McIntyre, Famiglietti had at least opened the way for the paper to finally recommence the peer review process.

The difficulties with the verification statistics were not the least of Wahl and Ammann's problems either. There had been a delay of nearly six months between Saiers' original rejection of the comment and his ousting by Famiglietti, so it was the end of November 2005 before the comment was back in play. Time was therefore running impossibly short for the paper to be included in the IPCC report: the first order draft had been issued and the review process was nearly complete. The Hockey Team therefore had only days left before they ran out of time, as the IPCC's timetable made absolutely clear:

> Third Lead Author meeting, December 13 to 15, Christchurch,

---

*There has been some confusion over what happened to Saiers, with some commentators believing that he lost his position at the journal entirely. In fact he only lost responsibility for dealing with McIntyre's paper, but remained at the journal for some time thereafter.

New Zealand. This meeting considers comments on the first order draft and writing of the second order draft starts immediately
afterwards. Meeting of the [Technical Summary and Summary for
Policymakers] writing team [on] December 16, [in] Christchurch,
New Zealand. Note. Literature to be cited will need to be published or in press by this time.*

The Second Order Draft was the last time the official reviewers would
see the report before it was published, so the deadline had presumably
been imposed to ensure that they would see and assess at least once
all the papers that the author team had cited. To emphasise the point,
the deadline had been made still more explicit in another document –
the 'Deadlines for Literature Cited in the Working Group I Fourth Assessment Report'. This was written by Martin Manning, the head of the
Technical Services Unit (TSU), the IPCC's administrative arm, and in it
Manning charged chapter authors with making sure that they only used
final versions of any paper they had considered during the writing of the
Second Order Draft, although he also made allowance for incorporating
copyediting changes after this time. Crucially, however, Manning also
explained what this would mean in practical terms:

> In practice this means that by December 2005, papers cited need
> to be either published or 'in press'.[34]

However, the 'in press' deadline was only half the story. As the 'Deadlines' document made clear, it was considered important that government
and expert reviewers only use versions of the papers that were complete
in all respects, and a second hurdle was therefore put in place:

> When the second draft of the [report] is sent to governments and
> experts for the second round review, the TSU must hold final pre-
> print copies of any unpublished papers that are cited in order that
> these can be made available to reviewers. This means that by late-
> February 2006 if [lead authors] can not assure us that a paper is
> in press and provide a preprint we will ask them to remove any
> reference to it.[34]

The stern warning that citations of papers that did not meet these
deadlines would be removed shows that the IPCC was taking compliance

---

*Quoted in David Holland's submission to the Russell review.[33] The document can
no longer be seen in its original web location.

with the timetable very seriously. Clearly there was no way that the *Climatic Change* paper was going to be 'in press' in two weeks' time and the added hurdle of having a preprint available by the end of February must have made the problem look almost insurmountable. The Hockey Team were, however, nothing if not resourceful and bypassing a mere deadline was to prove well within their capabilities.

BEATING THE CUT

The editor of *Climatic Change*, and therefore the man who was responsible for deciding the fate of Wahl and Ammann's paper, was Stephen Schneider, a climatologist from Stanford University in the USA. Schneider had been at the very centre of the global warming movement almost from the beginning and he was therefore completely trusted by the Hockey Team.* Their confidence was not misplaced; when he had received the first draft of Wahl and Ammann's paper back in May, Schneider had quite properly sent it out for review to McIntyre, with Phil Jones providing an opinion from the opposite end of the spectrum of opinion.† However, with time now of the essence, it appears that Schneider decided to tip the balance in favour of the Hockey Team: for the second draft of the paper McIntyre was not included among the peer reviewers.

With McIntyre out of the way, there would be no awkward questions about the contents of the paper, but there was still the problem of the IPCC deadline to deal with and, with time so short, Schneider was forced to go one step further. Although it is not known who came up with the idea, on the eve of the lead author meeting in Christchurch he introduced a new status for *Climatic Change* papers of 'provisionally acccepted'. There is no record of this status ever having been applied to other papers at the journal and it therefore appears that Schneider created it for the sole purpose of ensuring that the Wahl and Ammann paper could be cited in the IPCC report. However, he also appears to have been nervous about what he was getting himself into, and he picked up on some of McIntyre's criticisms of the paper, telling Wahl and Ammann that the verification statistics must be shown before he would clear it for publication. But 'provisional acceptance' was enough for Wahl, who sent an exultant email off to Overpeck and Briffa:

---

*He had, for example, been copied in on much of the correspondence relating to the Soon and Baliunas affair.
†Jones' friendly review of the paper was among the Climategate files.

I want you to know that we heard from Steve Schneider today that our paper with *Climatic Change* has been provisionally accepted for publication. The provisions Steve outlined are ones we fully accept and will implement (extra statistics of merit and remaking of graphics), so this paper can be viewed as accepted, I should think.[35]

But Wahl had some less favourable news too – the GRL comment had run into further problems. When Famiglietti had announced that the comment had been accepted, he had received a bitter complaint from McIntyre, who noted that Famiglietti had broken his own journal's rules in the way he had handled Wahl and Ammann's comment and that he had then compounded the problem by appearing to criticise McIntyre and McKitrick in a magazine interview. McIntyre said that through these actions Famiglietti appeared to be left hopelessly compromised as the editor responsible and asked him to hand over to someone more neutral. Famiglietti's position was difficult – his breach of the rules had been so transparent that it was likely to rebound badly on him, particularly as McIntyre would probably write an excoriating response to the comment. However, with Wahl and Ammann's *Climatic Change* paper moving safely into the IPCC process, it was now possible for Famiglietti to backtrack somewhat and save face: just days after apparently accepting the comment outright, he wrote to Ammann to explain that his comment on MM05 would actually have to run the gauntlet of peer review once again.[36]

THE SECOND DEADLINE

At the start of 2006, and with work under way on the Second Order Draft, Wahl wrote once again to Schneider and Briffa, this time enclosing the revised version of the *Climatic Change* paper:

I'm not sure that I ever sent you the updated Wahl–Ammann paper that was the basis for [Schneider's] provisional acceptance. Here it is. As is, it contains a long appendix. . . [on verification statistics], which was not in the version I had sent you earlier in the year. All the main results and conclusions are the same.[36]

Wahl went on to explain that Schneider and the peer reviewer of the second draft had requested further changes to the manuscript, involving the vexed question of the verification statistics, saying that he and Ammann were going to:

... address publishing [the R^2 and CE] calculations for verification, which Steve [Schneider] and the reviewer reason should be done to get the conversation off the topic of us choosing not to report these measures, and onto the science itself.[36]*

Wahl's message shows clearly that the version of the paper considered by the author team at their meeting before Christmas had clearly not been complete in all material respects as demanded by the Deadlines document – it had not included the adverse verification statistics, which would have shown that the conclusions were unreliable.[†]

What was worse for the author team, Schneider's concoction of a 'provisionally accepted' status had been observed by McIntyre, who had written a long blog post with a detailed analysis of the timings involved. This was potentially very serious for the author team's credibility, as Eystein Jansen explained to Overpeck:

> Hi Peck, I assume a provisional acceptance is OK by IPCC rules? The timing of these matters are being followed closely by McIntyre ... and we cannot afford to being caught doing anything that is not within the regulations.[36]

Overpeck was not optimistic, and appeared resigned to the Wahl and Ammann paper being non-compliant, at least on the first deadline. However, he seemed to be pinning his hopes on it getting full acceptance by the time of the second deadline at the end of February 2006: if Wahl and Ammann could come up with a preprint by then, it would presumably allow Overpeck to argue that the paper should be included in the report regardless of it having missed the earlier deadline back in December:

> I'm betting that 'provisional acceptance' is not good enough for inclusion in the Second Order Draft, but based on what Gene [Wahl] has said, he should have formal acceptance soon – we really need that. Can you give us a read on when you'll have it Gene? Best make this a top priority, or we'll have to leave your important work out of the chapter.[36]

---

*Note that the wording suggests that there was a single reviewer of the second draft of the paper.
†See p. 20.

At the start of February,* Wahl wrote to update the IPCC team on progress. He and Ammann were still working on the changes to the verification statistics that Schneider had demanded, but he suggested that this work could be complete within days and that Schneider might then be able to provide unconditional acceptance for the paper in a similarly short time.

Overpeck's reply suggested, however, that he had almost given up hope of the paper meeting the deadline:

> Based on your update (which is much appreciated), I'm not sure we'll be able to cite [the paper] in the [Second Order Draft]...The rule is that we can't cite any papers not in press by end of Feb.
>
> From what you are saying, there isn't much chance for in press by the end of the month? If this is not true, please let me, Keith, Tim and Eystein know, and make sure you send the in press doc as soon as it is officially in press (as in you have written confirmation). We have to be careful on these issues.[37]

Overpeck's statement is interesting because it appears to contradict the 'Deadlines' document, which said that not only must the paper be in press, but that the authors must be able to provide a preprint too. However, Overpeck's email appears to have given Wahl an idea, as he explained in his reply:

> ...as I have understood it in our communications with [Schneider], final acceptance is equivalent to being in press for *Climatic Change* because it is a 'journal of record'. However, this would need to be confirmed to be quite sure.[37]

Wahl then wrote to Schneider to seeking clarification:

> Overpeck...says that the paper needs to be in press by the end of February to be acceptable to be cited in the [Second Order Draft]. (I had thought that we had passed all chance for citation in the next IPCC report back in December, but [Overpeck] has made it known to me this is not so).
>
> He and I have communicated re: what 'in press' means for *Climatic Change*, and I agreed to contact you to have a clear definition. What I have understood from our conversations before is that if

---

*The timing is approximate since the date is unclear from the email thread. It may have been during late January.

you receive the [manuscript] and move it from 'provisionally accepted' status to 'accepted', then this can be considered in press, in light of [*Climatic Change*] being a journal of record.[38]

This email appears twice in the CRU disclosures – once in the message sent to Schneider and once when the correspondence was copied to Overpeck later that month.[39] Intriguingly, on the latter occasion, Wahl appears to have deleted the parenthetical sentence in the first paragraph about his having understood that the deadline had been missed.

Shortly afterwards Schneider replied, telling Wahl that his interpretation was 'fine', but he also emphasised the need to complete the revisions quickly so that the paper could get through its peer review in time for the IPCC. Despite all the problems with Wahl's two papers, it looked as they might just beat the deadline after all.

Finally, at the end of the month, and just hours before the deadline expired, Wahl wrote triumphantly to Overpeck:

> Good news this day. The Wahl–Ammann paper...has been given fully accepted status today by Stephen Schneider. I copy his affirmation of this below, and after that his remark from earlier this month regarding this status being equivalent to 'in press'.[40]

'Accepted' had been deemed to be equivalent to 'in press', the failure of Wahl and Ammann to provide a preprint of the paper had been overlooked, but at last the IPCC had got its rebuttal of McIntyre and McKitrick.

THE DRAFTS

With Wahl and Ammann's paper safely delivered into the IPCC process at the end of February 2006, Briffa could finally set about completing his drafting of the chapter. However, his task remained a difficult one; he was under considerable pressure to give a picture of global temperature history that he felt was misleading and that understated the uncertainties. As he told Overpeck and Jansen,

> ...we are having trouble to express the real message of the reconstructions – being scientifically sound in representing uncertainty, while still getting the crux of the information across clearly. It is not right to ignore uncertainty, but expressing this merely in an arbitrary way...allows the uncertainty to swamp the magnitude of the changes through time.

> ...you have to consider that since the [Third Assessment Report],
> there has been a lot of argument re [the] 'hockey stick' and the real
> independence of the inputs to most subsequent analyses is mini-
> mal. True, there have been many different techniques used...but
> the efficacy of these is still far from established. We should be
> careful not to push the conclusions beyond what we can securely
> justify – and this is not much other than a confirmation of the gen-
> eral conclusions of the [Third Assessment Report]. We must resist
> being pushed to present the results such that we will be accused of
> bias... [41]

Although Briffa's conscience appeared to be pushing him towards
recognising the problems with Hockey Stick and the other millennial temperature reconstructions in the report, he felt that some other members of
the IPCC team were much less scrupulous in their approach to the uncertainties; he was being pushed towards taking a much less even-handed
approach. As he explained, his concerns were centred around two people in particular: Mann and Susan Solomon, the scientist in charge of the
whole scientific report:

> Of course this discussion now needs to go to the wider chapter
> authorship, but do not let Susan (or Mike) push you (us) beyond
> where we know is right. [41]

To add to Briffa's problems, word arrived that the United States National Academy of Sciences (NAS), which at the time was investigating
the whole area of paleoclimate, [42] was going to provide little support for
the idea that proxy-based temperature reconstructions were reliable. The
news came via Richard Alley, a glaciologist from Penn State University,
who had made a presentation to the NAS inquiry and who was on cordial terms with several members of the Hockey Team. According to Alley,
the panel members had shown a great deal of interest in some of the
doubts raised over the reliability of the reconstructions. Alley was therefore worried that if the IPCC produced a report that appeared to place too
much confidence in the reconstructions, the contrast with the NAS panel's
conclusions would be an embarrassment to both groups.

The pressure on Briffa must have been enormous, with Alley and his
conscience suggesting caution and Mann and Solomon pushing him in
the other direction. The Second Order Draft is therefore a remarkable
document, partly because it had little of the caution that Alley had sug-

gested, but also when read in the light of the statements Briffa had made to Overpeck and Jansen.

Briffa's new assessment of the millennial temperature reconstructions stated that Wahl and Ammann had produced a replication of the Hockey Stick, succeeding where McIntyre and McKitrick had failed in MM03. However, this was hardly a full and fair description of what had happened, since at that time the two Canadians had written their first paper, Mann had been withholding key details of his methodology. It is therefore not surprising that the two Canadians had found that an exact replication eluded them. Briffa then went on to note the details of the new critique made by the two Canadians in MM05 – the verification statistics and the biased algorithm – but failed to make it clear where the argument now stood:

> The latter may have some foundation, but it is unclear whether it has a marked impact upon the final reconstruction.[43]

In fact, McIntyre and McKitrick had issued detailed refutations to each of the critics who had suggested that Mann's biased methodology had only a limited impact on the final reconstruction,[44] and indeed, just a few weeks later, Tim Osborn reported the opinions of a group of top paleoclimatologists on this very question...

> In general, most people accepted that the MBH method could, in some situations, result in biased reconstructions with too little low-frequency.[45]

...in other words there *was* a significant impact. But while Briffa had at least made some kind of statement about the methodological problems with the Hockey Stick, he stayed resolutely silent on the subject of the verification statistics. Did McIntyre's observation that Wahl and Ammann's version of the Hockey Stick failed its verification R^2 have any foundation? Briffa was not saying.

Lastly, and rather remarkably, Briffa said that because Mann had come up with a reconstruction that was broadly similar to those of other scientists working in the area, he had probably arrived at the correct answer regardless of any problems with his data and methods:

> However, subsequent work using different methods to those of Mann et al. (1998, 1999), also provides evidence of rapid 20th century warming compared to reconstructed temperatures in the preceding millennium.[43]

So while he had told Overpeck and Jansen that the efficacy of these different methods was 'far from proven', he was now suggesting to the public that they provided assurance of the soundness of the Hockey Stick. Moreover, the new draft contained not a hint of the caveat Briffa had delivered to Overpeck about the 'independent verifications' of Mann's work being nothing of the sort – they largely relied on the bristlecones, the same contaminated data that Mann had used. The message for public consumption was clearly going to be very different to the one communicated in private.

MORE DEADLINE PROBLEMS

Wahl may have thought his problems were over once his paper was accepted for the Second Order Draft, but in fact the respite was only temporary. Just two weeks after the passing of the IPCC deadline, McIntyre forwarded his response to the revised Wahl and Ammann comment to Famiglietti, who shortly afterwards announced that he, like Saiers before him, had decided that the comment was unpublishable. This was extraordinary – the journal had first rejected the submission out of hand, then had ousted the editor responsible for doing so, then had resurrected the comment and finally had decided to reject it once again. Exactly what was going through Famiglietti's mind is not clear, but it may have been that he had decided that it was simply too embarrassing to publish a paper that was so full of errors and misrepresentations – McIntyre had not been gentle in his comments on what Wahl and Ammann had done. Moreover, with Wahl and Ammann's *Climatic Change* paper now safely in the IPCC review process, Famiglietti may have felt that he had done his bit. So although rejecting the comment again left the paper in trouble once more – after all, it relied upon the comment for its statistical arguments – in practice it turned out to present no difficulties for the IPCC team, who simply carried on as if nothing had happened.

Further difficulties were emerging too. The Second Order Draft had been completed and sent out to reviewers at the end of March 2006, with the new text indicating that Wahl and Ammann had been able to replicate the Hockey Stick in full. However, the use of the paper to support the Hockey Stick was about to backfire. During May, the expert reviewers started to submit comments on the Second Order Draft, and Overpeck and his team will surely have been horrified to see that several commenters had pointed out that the version of Wahl and Ammann's paper on the IPCC website was different to the final draft that Ammann had posted on his own home page, demonstrating conclusively that the version considered by the author team was not the final one.[46] This could only mean one thing: that Wahl and Ammann's paper had failed to meet the deadline. For a while there was consternation among the author team as they tried to work out exactly which version of the paper they had been looking at when the deadline had been reached, although Overpeck declared that the differences were not material. However, it was becoming increasingly clear that the paper's inclusion was going to leave the author team exposed to a great deal of criticism.

As well as Wahl and Ammann's effort, there were several other papers supportive of the Hockey Team position that had come into print too late for the IPCC deadline, including one by Briffa himself. This was a problem for the author team: the IPCC's own rules were preventing them from incorporating these helpful findings. However, once again the Hockey Team proved to be more than capable of extricating themselves from their dilemma.

At the end of June 2006 the author teams were to meet to discuss the review comments on the Second Order Draft in Bergen, Norway. At that meeting Solomon and Manning, together with Trenberth, Jansen and the coordinating lead authors for the other chapters,* discussed the issue of compliance with the deadlines and it appears that they simply decided to rewrite the rules, as Manning explained in an email to the expert reviewers shortly afterwards:

> We are very grateful to the many reviewers of the second draft of the Working Group I contribution to the IPCC Fourth Assessment Report for suggestions received on issues of balance and citation of additional scientific literature. To ensure clarity and transparency in determining how such material might be included in the final Working Group I report, the following guidelines will be used by Lead Authors in considering such suggestions.

> In preparing the final draft of the IPCC Working Group I report, Lead Authors may include... papers published in 2006 where, in their judgment, doing so would advance the goal of achieving a balance of scientific views in addressing reviewer comments. However, new issues beyond those covered in the second order draft

---

*Also present was Solomon's co-chair at the head of Working Group I. Daihe Qin. The identities of those involved in taking the decision to change the deadline were revealed in an appendix to Osborn and Briffa's submission to the Russell review.[47]

will not be introduced at this stage in the preparation of the report.

Reviewers are invited to submit copies of additional papers that are either in-press or published in 2006...[48]

Manning's email was extraordinary on a number of levels. Firstly there is almost no evidence that reviewers of the Second Order Draft had been making suggestions about citation of new papers. In fact, there appears to have been only a single such request, and this had apparently been rejected, with the authors noting that they would only examine recent papers that met IPCC deadlines.[49] Moreover, because the Second Order Draft was the last time the official reviewers would see the report before its publication, the new citations and any new text involved would go completely unreviewed. The perception that Manning's email was an unofficial attempt to assist the author team in the battle over the Hockey Stick is hard to avoid.

This view is reinforced by a later discovery. The IPCC is an intergovernmental body and governments are therefore closely involved in its activities. In particular, the way the panel operates is determined by agreement among the governments, which means that any changes to the rules have to be agreed by the governments' representatives. As well as setting down the rules, governments provide their own review comments on the drafts of the report. The US government in particular had been one of those that had commented on the inconsistency of the different versions of the Wahl and Ammann paper – they at least were aware of the issues over the deadline and could therefore be expected to object to any changes to the rules to allow new citations to be added after the completion of the review. However, this eventuality had been foreseen and the possibility of one of the governments objecting was sidestepped by the simple expedient of sending the email advising of the rule changes to the expert reviewers alone. The governments were kept in the dark.*

SECRET COMMUNICATIONS

In the months that followed, Briffa's problems grew. As well being pressured by Mann and Solomon he was having to deal with his review editor, Professor John Mitchell of the UK's Meteorological Office. Mitchell, along

---

*David Holland ascertained this information by making an FOI request for the email to the UK government department responsible for dealings with the IPCC. He was informed that they did not have it.

with his colleague Jean Jouzel, was responsible for refereeing any disputes in the paleoclimate chapter and ensuring that both sides of any debate were represented. Having seen the review comments on the Second Order Draft Mitchell knew that the Hockey Stick debate was going to require his attention, and he wrote to Overpeck offering his advice, and copying Osborn and Briffa. It was important, he said, to give a 'clear answer to the skeptics' and he set out what he thought needed to be covered:

> Our response should consider all the issues for both [the Hockey Stick] and the overall chapter conclusions:
>
> a. The role of bristlecone pine data
>
>> Is it reliable?
>>
>> Is it necessary to include this data to arrive at the conclusion that recent warmth is unprecedented?
>
> b. Is the [principal components analysis] approach robust? Are the results statistically significant? It seems to me that in the case of [the Hockey Stick] the answer in each is no. It is not clear how robust and significant the more recent approaches are. [50]

So clearly, Mitchell was in full agreement with Briffa's position on the Hockey Stick and the 'independent confirmations':[*] none of them were reliable. In these circumstances, a clear response to the sceptics that also kept Solomon and Mann happy looked as though it would be almost impossible to achieve.

With the deadline for inclusion in the report redrawn, Wahl and Ammann's paper and all the papers that the author team thought would support the IPCC consensus could be incorporated into the text. However, this approach had an unfortunate downside in that sceptics among the reviewers would now be able to suggest their own new papers too. McIntyre was fully aware of this opportunity and decided to submit the reports by two separate inquiries into the temperature reconstructions: the NAS report[†] into the field of paleoclimate and the Wegman report into the statistics of the Hockey Stick. These two inquiries had confirmed many of McIntyre and McKitrick's criticisms and the Wegman report, in particular, was quite unequivocal in its support for McIntyre's statistical

---

[*]See p. 25.
[†]See p. 26.

criticisms of the Hockey Stick, describing them as 'valid and compelling'. However, the NAS panel had tempered its support with an observation that was similar to the text in Briffa's latest draft – namely that despite using inappropriate data and a biased algorithm, Mann's findings appeared to be supported by other, allegedly independent, temperature reconstructions.

The mention of the findings of either the Wegman or NAS panels in the final report could potentially have been disastrous for the IPCC. Once again, the steps taken to deal with the problem appear to have been unethical. Shortly after the new deadline expired, the Technical Services Unit (TSU) issued a spreadsheet to the author teams which listed the new papers that had been suggested by the authors and reviewers. It has been determined, however, that this did not mention either the Wegman or NAS reports.* Once again, the scales were being tipped decisively in favour of the Hockey Team's position. What is more, Briffa was about to tip them even further.

The previous summer, shortly after Wahl and Ammann's announcement of their two journal submissions, Wahl had asked Overpeck if he and the IPCC team would like a copy of the manuscript of the GRL comment to help them in their work. This suggestion was somewhat unorthodox: there is an official register of IPCC reviewers and Wahl's name was not on it. On this occasion, Overpeck had replied enthusiastically, saying that Wahl should send it over if he was comfortable doing so.[30] However, Overpeck's reaction to the offer is remarkable when set against the way he dealt with Neil Roberts, a scientist from the University of Plymouth who some time later made a similar offer of informal assistance with the review. On that occasion, Overpeck had explained in no uncertain terms that it was necessary to be registered as an official reviewer before contributing:

> Since the IPCC has very strict rules about all this, I'm going to ask them. . . to send you an official invitation to review, along with the process – formal, but highly efficient - to follow. If you could send your comments in that way it would be a great help. We've been asked to keep everything squeaky clean, and not to get comments informally.[51]

Of course, the inconsistent treatment of the two offers of informal help may have been prompted by the discovery that McIntyre was watch-

---

*See below, p. 198.

ing them so closely or perhaps by a directive from further up the IPCC organisation. However, since Overpeck had copied this last message to Briffa, it is clear that by May 2006 Briffa was fully aware that further correspondence with unregistered reviewers was forbidden. Despite this, the pressure of dealing with the Hockey Stick war appears to have been too much: just weeks later, he wrote to Wahl:

> Gene
>
> I am taking the liberty (confidentially) to send you a copy of the reviewers comments (please keep these to yourself) of the last IPCC draft chapter. I am concerned that I am not as objective as perhaps I should be and would appreciate your take on the comments. . . that relate to your reassessment of the Mann *et al.* work. I have to consider whether the current text is fair or whether I should change things in the light of the sceptic comments. . . I must put on record responses to these comments – any confidential help, opinions are appreciated.[52]

The fact that Briffa requested confidentiality suggests that he knew that he was breaching the rules. What is more remarkable still is that while Wahl, who was not an official IPCC reviewer, was being sent a copy of the review comments, offical reviewers, including McIntyre, were being told that they would have to visit the official IPCC archive in Harvard, Massachusetts if they wanted to get a look at them.[53]

Wahl was happy to help and a few days later he provided detailed input, sending over a copy of his GRL comment on MM05 to help Briffa's deliberations. But he too appeared to be quite clear that the course of action he and Briffa were embarking on was illegitimate, although his concern may have been more to do with the possibility of the journal discovering that he had breached the confidentiality of the peer review process than any breach of IPCC rules.

> Please note that [the comment] text is sent strictly confidentially – it should not be cited or mentioned in any form, and MUST not be transmitted without permission. However, I am more than happy to send it for your use, because it succinctly summarizes what we have found on all the issues that have come up re [The Hockey Stick].[52]

Interestingly, along with the text of the paper Wahl enclosed a summary of the issues surrounding the Hockey Stick that he had written for

the benefit of what he called 'a person in [Washington] DC' who was involved with the hearings about Mann's paper that were to be held in the
wake of the Wegman and NAS reports. The political importance of the
Hockey Stick was apparently undiminished.

Over the next few days emails shuttled back and forth between Wahl
and Briffa, with Briffa seeking comments on what he had written in the
text and in the responses to reviewers and Wahl putting forward his
own suggestions. Throughout, the two men were focused on the way
to deal with the Hockey Stick and how to handle McIntyre's criticisms.
And throughout they were at pains to ensure that no hint of what had
happened should reach the outside world:

> What I am concerned about for the time being is that nothing in
> the [comment] shows up anywhere...
>
> ...Please do not pass these on to anyone at all...
>
> ...PLEASE REMEMBER that this is 'for your eyes only'... [54]

Wahl's ideas were apparently proving useful, and Briffa thanked him
for his help and asked once again for reassurance that no details of their
correspondence would be revealed:

> I have 'borrowed (stolen)' from two of your responses in a signifi-
> cant degree – please assure me that this OK (and will not later be
> obvious) hopefully. [54]

Briffa spoke of the stress he was working under and Wahl only made
things worse when he explained that publication of his paper might not
even take place in 2006, which would put him in breach of even the new
deadline.

> I should note that [the paper] is still in 'in press' status,* and its
> exact publication date will be affected by publication of an edito-
> rial designed to go with it that Caspar and I are submitting this
> weekend. Thus I cannot say it is certain this article will come out
> in 2006, but its final acceptance for publication as of [28 February
> 2006] remains completely solid. [54]

---

*Wahl refers to 'AW 2006', which might seem to suggest that he was talking about
the comment rather than the paper. However his reference to the acceptance date in
February makes it clear that he is referring to the paper.

THE REPORT

The final version of the IPCC report appeared in early 2007 and it was very much as the sceptics had feared.[55] In the Second Order Draft, Briffa had said that the effect of McIntyre's criticisms was unclear. Now, however, he was more definite, claiming that Wahl and Ammann's paper had shown that the impact of the errors McIntyre had identified was 'very small'. Moreover, as expected, Briffa noted that there were other reconstructions around that provided support for Mann's general conclusions. The verification statistics were mentioned in passing but a definitive statement was sidestepped. Of the bristlecones there was not a word.

Wahl and Ammann's paper was eventually published in September that same year, more than eighteen months after reaching 'in press' status, demonstrating fairly conclusively that it had failed to meet the deadline that had been moved for its benefit. It was accompanied by another paper: the comment on McIntyre and McKitrick's work, which Famiglietti had apparently refused to publish in GRL. Wahl and Amman had therefore rewritten it and submitted it to *Climatic Change*; having agreed to accept the paper despite it relying on the unpublished comment, editor-in-chief Stephen Schneider was put in a nearly impossible position by GRL's rejection of the comment for a second time – he simply had to offer *Climatic Change* as an alternative outlet for the comment, despite all its flaws. The story of the scientific distortions that Schneider was forced to accept in the process has been told elsewhere.[53]

THE PURSUIT

The Hockey Team may have felt that they had won an important victory, but the sceptics were only just getting into their stride. As McIntyre unearthed each new detail of the manoeuvrings behind the scenes at the IPCC, he set out what he had discovered for all to read at his *Climate Audit* blog. Since he had set it up in 2005 the site had turned into a hive of activity, and McIntyre now had a loyal band of followers who were increasingly willing to help him out with his growing workload. One of the readers who took some of the strain was David Holland, a retired electrical engineer from England. Holland had been a regular commenter at *Climate Audit* for several years and had been heavily involved in sceptic efforts in the UK, although he still kept a relatively low profile. This, however, was soon to change.

Holland's first target was to understand how the IPCC review process had worked in practice. McIntyre and McKitrick had submitted many detailed comments on the paleoclimate chapter and had been repeatedly fobbed off with unresponsive replies from Briffa. This dismissal should have been picked up by the chapter review editors, Mitchell and Jouzel, whose job was to ensure that scientific disputes were correctly represented in the final report. However, despite the IPCC's rules requiring all review comments to be made public, the contributions of Mitchell and Jouzel remained unpublished. Holland therefore decided to force the issue, requesting copies of Mitchell's comments under UK freedom of information (FOI) laws.

At first Mitchell resisted the request, claiming unconvincingly that he did not have a copy of the report that he had written. He suggested that Holland ask the TSU to send it to him, although he claimed, equally unconvincingly, that he did not have their email address. Fortunately, the IPCC appear to have overruled Mitchell and by the start of 2008 Holland was in possession of the full set of review editor comments. These, however, turned out to be a huge disappointment. Instead of a description of the disputes in each chapter and how they had been represented in the report, each one amounted to a simple sign-off by the review editor, affirming that they had reviewed their chapter and were happy with it. The sole exception to this pattern turned out to be from Mitchell, who had discussed the Hockey Stick:

> As Review Editor of Chapter 6... I can confirm that the authors have in my view dealt with reviewers' comments to the extent that can reasonably be expected. There will inevitably remain some disagreement on how they have dealt with reconstructions of the last 1000 years and there is further work to be done here in the future, but in my judgment, the authors have made a reasonable assessment of the evidence they have to hand. The other possible area of contention (within the author team) is on some aspects of sea-level rise. This has gone some way towards reconciliation but I sense not everyone is entirely happy.
>
> With these caveats I am happy to sign off the chapter... [56]

So instead of discussing how Briffa had achieved what was required of him – to fairly represent both sides of the Hockey Stick dispute in the report – Mitchell had accepted that Briffa could act as judge and jury,* a

---

*Note that Mitchell seems to have signed the report with a date of 8 December 2007. This is nine months after the publication of the report. At some point subsequently, the date has been amended, apparently in another hand, to 2006.

decision that appeared to give the lead authors carte blanche to ignore the IPCC's procedures.

The brevity of the review looked very troubling, and Holland therefore decided to dig a little further – it occurred to him that there might be more to Mitchell's review than simply the sign-off. Shortly afterwards he issued a further FOI request to the Met Office, asking for copies of any working papers produced by Mitchell during his work on the IPCC report. A few days later he extended his inquiry again, sending a similar request to Sir Brian Hoskins of the University of Reading, who had been a review editor on another chapter of the report.

Holland's FOI requests seemed to have caused a measure of consternation among their recipients. Some days after receiving Holland's request, Mitchell told Susan Solomon what was happening and asked how he should proceed, copying the message to Jouzel and Briffa, as well as to the TSU, whose email address he had told Holland he did not have just a few weeks earlier:

> I have received the following letter from David Holland, who has links with Stephen McIntyre and his Climate Audit website, on the review process for chapter 6 of [the Fourth Assessment Report]. I have discussed this briefly with Jean [Jouzel] and we do not think there is an issue. However given the wider nature of the questions, I think it would be more appropriate for any response to come through IPCC rather than me as an individual. I will wait to hear from IPCC before I respond...I understand [Sir Brian Hoskins] has received a similar enquiry, hence I have included his name on the copy list.[57]

Solomon was adamant that nothing should be released, informing Mitchell that the proper sources for people who wanted to understand the review were the reviewers' comments and the authors' responses to them. She explained that it would be 'inappropriate to provide more information'.[57] This was a remarkable position for her to take, since she must have been aware that Holland had made a formal, legally binding request for information – she was in essence asking Mitchell to break UK FOI legislation. And it was not just Mitchell; her reply was copied to Briffa and to key figures in the IPCC, including Renate Christ, the organisation's secretary, and all 27 review editors. Everyone needed to know that the sceptics were hot on their trail and that no disclosures should be made.

When Mitchell replied to Holland's letter he was once again evasive, echoing almost to the letter Solomon's request to stonewall, and going on to claim that he had not kept any of his working papers.

> There is no requirement to do so, given the extensive documentation already available from IPCC. The crux of the review editors' work is carried out at the lead authors meetings going through the chapters comment by comment with the lead authors. [58]

Mitchell's suggestion that he did not have to keep his working papers is not correct, since IPCC procedures require that all expert review comments are retained as part of the public record. His reply therefore seemed so unlikely that Holland decided to keep pressing, and at the start of April he issued another FOI request, this time asking for all of Mitchell's email correspondence in connection with the Fourth Assessment Report. Once again the Met Office's response was remarkable; although they released correspondence that post-dated Holland's first FOI request, they claimed that they held no relevant information from earlier dates. This implied one of two extraordinary scenarios – either Mitchell had not sent or received any emails in connection with the IPCC report or he had destroyed all this correspondence within months of the report's publication. And as we have seen, there *had* been some correspondence: Mitchell had emailed Overpeck and Jansen in the closing stages of the IPCC review to discuss how the Hockey Stick affair should be handled.*

Although the Met Office may have felt that they had successfully complied with Solomon's request not to reveal any further details of the review process, the information they did release was very interesting to Holland. Among the disclosures were Mitchell's email to Solomon discussing his FOI request and Solomon's instruction to the scientists involved in the review process to reveal nothing. This showed Holland clearly that the IPCC was mounting a stubborn resistance, and he determined to redouble his efforts.

The next step was a letter to Briffa, probing several of the outstanding questions about the Fourth Assessment Report. Like Mitchell, Briffa decided not to answer immediately, writing to tell Overpeck and Jansen that he would reply when he got round to it. Holland, however, was still considering ways to get a glimpse of what was happening behind the IPCC's facade, and he soon struck upon a different way of tackling the problem.

---

*See p. 31.

He could also see that, as well as Mitchell and Briffa, several other UK-based scientists had received Solomon's email and it was possible that some of the institutions at which they worked might be more forthcoming with information. He therefore decided to extend his requests. First, he formalised the message to Briffa into an FOI request, asking for all of Briffa's IPCC-related correspondence. At the same time he sent further requests to DEFRA, the government department that co-ordinated the UK's involvement in the IPCC, and to the universities of Oxford and Reading, who had also provided review editors to the IPCC.*

At CRU, Holland's request seems to have caused some concern and just four days after Holland had issued his request, Jones emailed Mann, Bradley and Ammann to tell them what was happening.

> You can delete this attachment if you want. Keep this quiet also, but this is the person who is putting in FOI requests for all emails Keith and Tim [Osborn] have written and received re [the paleo-climate chapter of the Fourth Assessment Report]. We think we've found a way around this.[59]

THE DOG THAT DIDN'T BARK

While Holland was digging away, trying to discover what had happened behind the scenes in the final stages of the IPCC review, McIntyre had been pondering the responses given to his comments on the Second Order Draft and had stumbled across something rather remarkable. It had begun when he noticed that Ammann appeared to have failed to submit any comments to the IPCC review process, something that seemed unlikely given his close involvement in the Hockey Stick debate. Then, in mid-2007, the comment on MM05 that GRL had twice rejected was finally published in *Climatic Change*, and McIntyre quickly noticed something else that was very odd: some of his review comments on the Second Order Draft had been rejected by Briffa using arguments and turns of phrase that bore an uncanny resemblance to language in the new Wahl and Ammann comment. The problem with this was that the comment had not even been *submitted* to *Climatic Change* until well after the deadline for submission of papers to the IPCC review. McIntyre quickly surmised the truth: Briffa had been taking advice on how to deal with McIntyre's arguments outwith the IPCC process. At first he assumed, incorrectly, that

---

*Myles Allen and Sir Brian Hoskins had been review editors on Chapters 10 and 3 of the report, respectively.

this advice had been provided by Ammann, the lead author of the comment – as we have seen it was actually Wahl who had been the source. However, no matter who had provided the information, it was clear that Briffa had been guilty of multiple breaches of the IPCC's rules, which required reviews to be open and transparent, and the literature cited to be peer reviewed and in print. Neither Wahl nor Ammann were official IPCC reviewers, so all this chicanery had taken place entirely outwith official channels. McKitrick later explained why this mattered:

> The problem, apparently, was that the actual publication record was either over [Briffa's] head or yielded a message he was disinclined to report, or both. So he went outside the structure of the IPCC report-writing process to recruit a highly partisan coach ('Gene') to provide him some text which would not be shown to the expert reviewers but which would go right into the final draft and be represented as the result of the official IPCC report-writing process. [60]

His trickery was now exposed, but things became even worse for Briffa the following day, when McIntyre revealed that he had also noticed that the IPCC had rewritten the timetable for submission of papers to the Fourth Assessment Report so as to allow time for Wahl and Ammann to finish their paper. [61] All the breaking and bending of rules looked as though it was going to come back to haunt the Hockey Team.

The sceptic community was abuzz with these new findings and Holland wasted no time in acting on them, sending out further FOI requests to CRU, this time specifying that he wanted any correspondence relating to the Wahl and Ammann paper or to changes in the timetable for the IPCC report.

The pressure was building inexorably at CRU and shortly afterwards David Palmer, the university's FOI officer, emailed Jones, Briffa and Osborn, telling them about Holland's new request. He said that he wanted to respond 'by the book', and noted that a refusal was likely to end in Holland appealing to the Information Commissioner.*

The scientists must have realised that their situation was perilous and they appear to have studied the legislation long and hard, looking for some way in which they could reject Holland's request. Before long they found what they thought was their get-out clause, in the shape of an

---

*The Information Commissioner is the official charged with enforcing FOI legislation in the UK.

exemption for information provided in confidence. There was, however, a potential problem in this line of reasoning in that the IPCC rules stated clearly that the organisation's overriding principles were of openness and transparency. But there were no other options on the table and just hours after Holland had sent his request, Osborn wrote to Caspar Ammann:

> Our university has received a request, under the UK Freedom of Information law, from someone called David Holland for emails or other documents that you may have sent to us that discuss any matters related to the IPCC assessment process.
>
> ... it would be useful to know your opinion on this matter. In particular, we would like to know whether you consider any emails that you sent to us as confidential.[62]

If the hint to Ammann was not dubious enough, the following day, Jones emailed Palmer and a senior UEA faculty manager named Michael McGarvie, to discuss how to resist disclosure.

> Keith (or you Dave) could say that... Keith didn't get any additional comments in the drafts other than those supplied by IPCC... [he] should say that he didn't get any papers through the IPCC process either.
>
> ... What we did get were papers sent to us directly – so not through IPCC, asking us to refer to them in the IPCC chapters. If only Holland knew how the process really worked!![63]

This is a remarkable document, which demonstrates that senior staff at UEA conspired to breach FOI laws. However, what followed was even more extraordinary. The next day, Jones started to take steps to ensure that there was no way Holland would be able to find the truth elsewhere, asking Mann to delete his correspondence with Briffa:

> Can you delete any emails you may have had with Keith re [the Fourth Assessment Report]? Keith will do likewise... Can you also email Gene [Wahl] and get him to do the same? I don't have his new email address. We will be getting Caspar [Ammann] to do likewise.[64]

The message is particularly damning because, as its title 'IPCC & FOI' made clear, Jones was aware of the relevance of the FOI laws and knew that they had a direct bearing on what he was asking Mann to do.*

---

*The title of Jones' message is surmised from Mann's response, which has the subject, 'Re: IPCC & FOI'.

Mann's reply also made it quite clear that he intended to comply with Jones' request:

> I'll contact Gene about this ASAP.[64]

Shortly afterwards, Mann forwarded Jones' request, apparently without comment, and Wahl dutifully deleted the emails.[65]

There was still the question of Ammann's correspondence, however, and it required a reminder from Osborn to prompt a response, with Ammann making a slightly more definite statement, saying that he might have written the emails differently had he known they were going to be made public. This, apparently, was good enough for the university and a few days later Holland's request was rejected on the grounds that the messages requested were confidential. Similar rejections followed from Oxford, Reading and DEFRA. The door had been slammed shut.

PRECAUTIONARY PRINCIPLE

Holland may have been seen off, but the sceptic blogs were still alive with the stories of the shenanigans around the Fourth Assessment Report and it must have been clear to everyone that more FOI requests would soon be issued. And with so many hard-to-defend decisions taken in recent months, it was imperative for the IPCC scientists that any such requests not be successful: precautionary measures were required.

Shortly afterwards, Jones explained what these measures had entailed in an email to Jean Palutikof, who until 2004 had been his co-director at CRU and who had been a senior figure in the IPCC.

> Jean
>
> ...What Keith and Tim did was to email all the [authors] on [the paleoclimate chapter], to ask if they would be happy for Keith/Tim to send emails relating to [their] discussions. They all refused, hence the refusal letter.
>
> ...John Mitchell did respond to a request from Holland. John had conveniently lost many emails, but he did reply with a few. Keith and Tim have moved all their emails from all the named people off their PCs and they are all on a memory stick.
>
> So any thoughts on how to respond?
>
> ...As you and Tom know Keith and I are nowhere near the world's best for structured archiving – working as we do on sedimentary sequencing!

Cheers

Phil[66]

Jones may well have believed that by simply moving emails from computers to memory sticks he could simply refuse the next FOI request from McIntyre or Holland, telling them that the information was not held. If so, he was almost certainly mistaken. When the FOI laws were framed, the possibility of public bodies trying to avoid compliance in this way was foreseen and the laws were written in such a way that the information on Jones' memory stick would be deemed to be held on behalf of the university. But more seriously, if a future FOI request had been refused, then a criminal breach of the legislation would almost certainly be committed in the process. The decision by the CRU scientists to hide their emails in this way was therefore a fateful one. And in time it would come back to haunt them.

CHAPTER 2

OF THERMOMETERS AND TREE RINGS

THE DATA

It was not only information about tree rings and the IPCC review that interested sceptics. One of the most important areas of research conducted at CRU was the instrumental temperature records – data from the network of weather stations used to record temperatures since the nineteenth century. Phil Jones had two main claims to fame as a climatologist, both of which were grounded in this area. Firstly he had published an important paper on the urban heat island effect (UHI) – the part of the measured warming that is due to waste heat from growing urban environments and is therefore non-climatic. Secondly, he was also responsible for combining the station records into a dataset called CRUTEM. This processed data was sent to the Met Office, where it was combined with ocean temperature records to produce the HADCRUT global temperature index, which was used prominently in the IPCC reports. The importance of CRUTEM and the UHI adjustment to the overall global warming hypothesis had made Jones the focus of considerable interest among sceptics, who had been trying to get hold of his underlying temperature station data for many years.

As far back as 2005, an Australian sceptic named Warwick Hughes had made repeated requests for the UHI data – first direct to Jones, then to the World Meteorological Organisation (WMO), and then to Jones again, but all to no avail. Hughes was nothing if not persistent, however, and although Jones had responded cordially to the earliest of Hughes' enquiries, he was eventually moved to reject the sceptic in the following terms, which were soon to make him a figure of some notoriety:

> Even if WMO agrees, I will still not pass on the data. We have 25 or so years invested in the work. Why should I make the data available to you, when your aim is to try and find something wrong with it? [67]

Hughes' requests eventually proved fruitless, but unbeknown to him the situation was about to change decisively in the sceptics' favour. In 2000, the incoming Labour government had introduced the UK's first freedom of information (FOI) legislation, although the new law was only to take force in January 2005, to give the bureaucracy time to prepare for the changes in working practices that would be required. So when Warwick Hughes made his final fruitless request to Jones in February 2005, it was actually already open to him to force Jones' hand by making an FOI request under the new law. Unfortunately Hughes, along with many residents of the UK, was unaware of this possibility and he gave up the chase for the time being.

Within CRU, however, there was considerable concern about these changes in the legal landscape. Shortly after the new law came into force, Jones received an email from Tom Wigley, the former director of the unit whom we met in the last chapter. Wigley said that he had heard that his work might now be subject to disclosure and asked if it was true. He was worried not only about his own data but also the work of colleagues from his days at CRU. These concerns appear to have been chiefly over computer codes, and Jones advised him (incorrectly) that since he was no longer an employee these would not be disclosable. Later on Jones also suggested that intellectual property considerations might provide an excuse not to disclose any information: as Jones put it, he would be 'hiding behind' the legal agreements made by CRU with the national meteorological services that supplied the raw data behind CRUTEM in the first place.

> I wouldn't worry about the code. If FOIA does ever get used by anyone, there [are also intellectual property rights] to consider as well. Data is covered by all the agreements we sign with people, so I will be hiding behind them. I'll be passing any requests onto the person at UEA who has been given a post to deal with them.[68]

A few days after his email exchange with Wigley, Jones also raised the question of FOI with Mann. Jones had just sent the raw station data to Mann's assistant, Scott Rutherford, and, mindful of a previous incident in which a sceptic had obtained data left on a publically accessible CRU server, he wanted to warn Mann to take care with the vital information.

> Just sent loads of station data to Scott...don't leave stuff lying around on FTP sites – you never know who is trawling them. [McIntyre and McKitrick] have been after the CRU station data for years. If they ever hear there is a Freedom of Information Act now in the UK, I think I'll delete the file rather than send to anyone...Tom Wigley has sent me a worried email when he heard about it – thought people could ask him for his model code.[69]

However, contrary to the expectations of Jones, Wigley and Mann, no FOI requests materialised in 2005, and by the the middle of the following year they must have started to relax somewhat. However, it was only a matter of time before someone in the sceptic community noticed the new law, and in the autumn of 2006, the moment that the CRU scientists had dreaded finally arrived. Willis Eschenbach, a sceptic based in Fiji, was a long-standing contributor at *Climate Audit* who for several years had followed the saga of the various attempts to extract data from Jones. Eschenbach's interest was in the CRUTEM temperature data and, learning about the introduction of the FOI Act, he realised that it could be important in bringing the long-running struggle to a decisive end. At the end of September 2006 he issued a request to CRU asking for the list of stations used in the CRUTEM temperature average and the raw data for each one.

Jones had been considering how to respond to such an FOI request for more than eighteen months and the story of the UEA's responses to Eschenbach's request suggests that this time had not been wasted. In their first reply, the university advised Eschenbach that the information he wanted was in one of two large international repositories of climate data: GHCN and NCAR. However, they failed to disclose the list of stations used in CRUTEM, so in fact their response represented a blatant case of stonewalling – the GHCN and NCAR archives hold temperature records for thousands of weather stations so it would be impossible for Eschenbach to replicate CRUTEM without knowing which of them had been used.

When Eschenbach protested, the university blanked him again, invoking the intellectual property approach that Jones had discussed with Tom Wigley the previous year: they claimed that some of the data was covered by confidentiality agreements with the national meteorological services that had supplied the data to CRU. Further correspondence was exchanged backwards and forwards between Palmer and Eschenbach, with UEA consistently refusing to address the request for a list of stations. Finally, however, they relented, agreeing to disclose this simple but vital piece of information. Even then there were complications. According to Palmer, and rather extraordinarily, the university did not have a list consisting solely of the sites currently used. It was six more months before Eschenbach received a list, and even then it was incomplete and error-ridden.

At the start of 2007, McIntyre decided to make use of the new FOI law himself, and issued a request for the underlying data for Jones' UHI paper. McIntyre had first developed an interest in UHI even before he began to look at the Hockey Stick. Right back in 2002 he had approached Jones for the underlying data, but at that point Jones had sent a vague reply about the data being on some discs which he was unable to locate. At the time McIntyre had chosen not to pursue the issue and had moved on to other things. However, his interest had been rekindled when he saw the UHI paper being cited again in the IPCC's Fourth Assessment Report. With his own involvement in the review now completed, and no major developments expected on the paleoclimate front, McIntyre decided that the time was ripe to take another look at Jones' most important contribution to the climate literature.

Jones' approach to assessing the affect of UHI on the overall global temperature trend was to compare the warming trends in urban temperature stations to those in rural ones – the difference between the two should in essence be the UHI effect. He had used networks of weather stations in, amongst other places, Australia, Russia and China, and concluded that the urban heat island effect was vanishingly small. In order for Jones' conclusions to hold, he had had to choose weather stations that were 'homogenous', which is to say they had not been moved from one place to another or suffered from other changes in their operation over the years. In his paper, Jones had noted that he and his co-authors had chosen sites 'with few, if any changes in instrumentation, location or observation times'.[70]

On the face of it, what Jones had achieved was extremely unlikely: weather station records even in the developed world were rarely homogenous. As McIntyre put it, 'the idea that China between 1954 and 1983 – the age of Chairman Mao and the Great Leap Forward – could have achieved consistency in temperature measurement that eluded the US observing system... is a conceit that seems absurd...'[71]

Some diligent research confirmed McIntyre's suspicions. According to a technical report about the quality of weather station data from China, there were actually very few homogenous station records in China, and indeed for the vast majority there were no station histories that would permit an assessment of the question in the first place. As we have seen, Jones had been unwilling to identify which stations he had used in the paper, but examination of the crude map included in his paper suggested that he had used many stations where the histories were not extant. The conclusion was unavoidable – he had not carried out the quality control procedures he claimed he had.

That was where things stood at the start of 2007. However, when Eschenbach revealed that he had requested the CRUTEM data under FOI, McIntyre needed little prompting to follow suit. If he could get hold of the list of stations used in Jones' 1990 paper he would be able to demonstrate conclusively that the relevant station histories did not exist. Shortly afterwards, McIntyre issued his first FOI request.

Perhaps unsurprisingly, UEA chose to handle McIntyre's request in exactly the same way as they did Eschenbach's, pointing the Canadian to the same pair of data repositories. However, McIntyre now had a different option open. Jones' paper had been published in the journal *Nature*, which has a policy of requiring its authors to make materials available to interested third parties on request. McIntyre therefore issued a materials complaint to the journal, and this approach finally did the trick. Shortly afterwards Jones relented and released the list.

KEENAN'S FRAUD ALLEGATION

There was great excitement among the *Climate Audit* regulars when the list of stations finally made its appearance in April 2007 – years of struggle to see the numbers behind this critical paper were at last over. Among the first sceptics to examine the figures was Doug Keenan, an independent researcher based in London. After a career in the City of London, Keenan now spent much of his time pursuing an interest in scientific misconduct. His work in the financial sector had been based around mathematics and statistics, and he had several publications in scientific journals to his name, in areas as diverse as radiocarbon dating and paleoclimate. He was therefore more than capable of understanding the relatively straightforward mathematics in Jones' UHI paper.

When the publication of the list of stations was announced Keenan set to work and it was only a few hours before he confirmed McIntyre's suspicions. Where the histories were extant, many of the stations appeared to have moved. Most of them, however, had no histories at all. What was worse, the list of stations confirmed the long-held suspicion that Jones' definition of rural stations stretched credibility to breaking point, with any settlement with a population less than 200,000 deemed acceptable as 'rural'. As we have seen, these were no narrow academic points – Jones' results were only credible if his claim of few station moves held up and if the rural stations really were rural; without these crucial features his results would have to be thrown out as unreliable.

However, there was another layer of complexity to the story that was quickly revealed. The Chinese data used by Jones had been provided by one of his co-authors, Wei-Chyung Wang, a researcher at the State University of New York, who had published a paper on Chinese temperature history at around the same time as Jones' UHI paper.* Although Jones had been one of Wang's co-authors, Keenan realised that his involvement in Wang's paper may have been quite limited: he may only have provided methodological advice for example, with Wang reciprocating by providing the Chinese data for Jones' own paper. It was therefore quite possible that Jones knew nothing of the problems with the station histories: the only person who must have known was Wang. So, after a week of checking his conclusions and having failed to get a response from Wang, Keenan wrote a letter that must have been devastating for its recipient.

> Dear Dr. Wang
>
> Regarding the Chinese meteorological data... it now seems clear that there are severe problems. In particular, the data was obtained from 84 meteorological stations that can be classified as follows.
>
> > 49 have no histories
> >
> > 08 have inconsistent histories
> >
> > 18 have substantial relocations
> >
> > 02 have single-year relocations
> >
> > 07 have no relocations
>
> Furthermore, some of the relocations are very distant – over 20 km. Others are to greatly different environments...
>
> The above contradicts the published claim to have considered the histories of the stations, especially for the 49 stations that have no histories. Yet the claim is crucial for the research conclusions.

---

*Wang obtained the data from another Chinese scholar, Zhaomei Zeng, who was rewarded with co-authorship of Wang's paper. However, Wang appears to have wanted to keep as much prestige as possible for himself, and Zeng's name did not appear on the Jones paper, despite the fact that she had prepared the data.

I e-mailed you about this on April 11th. I also phoned you on April 13th: you said that you were in a meeting and would get back to me. I have received no response.

I ask you to retract your... paper, in full, and to retract the claims made in *Nature* about the Chinese data. If you do not do so, I intend to publicly submit an allegation of research misconduct to your university... [72]

Keenan had copied Jones in on this message and the UEA man was clearly feeling the pressure, with Eschenbach's request for his CRUTEM data still a problem and his release of some of the UHI information looking as though it was going to make things worse rather than better. In an email to Trenberth and Mann, Jones wondered ominously if he was a 'marked man'. Mann was predictably belligerent, suggesting that Wang should sue Keenan for defamation, but Trenberth suggested a more underhand response:

...the response should try to somehow label these guys [as] lazy and incompetent and unable to do the huge amount of work it takes to construct such a database... my feeble suggestion is to indeed cast aspersions on their motives and throw in some counter rhetoric. Labeling them as lazy with nothing better to do seems like a good thing to do. [72]

More forthright support came from Ben Santer, a climatologist from the USA.

I looked at some of the stuff on the *Climate Audit* web site. I'd really like to talk to a few of these 'auditors' in a dark alley. [73]

As we have seen, there is no evidence that Jones knew of the problems with the Chinese data at the time of his paper's publication in 1990. However, McIntyre discovered that ten years later Jones had been coauthor on another paper that considered the issue of Chinese station histories and drew conclusions based upon the fact that there had been many station moves. [74] So by the turn of the millennium Jones must have been aware that his paper was flawed. Yet despite this, Jones continued to cite the 1990 paper, including in the IPCC's Fourth Assessment Report. This new discovery prompted further messages to Jones, with McIntyre suggesting that Jones should issue a correction and Keenan asking for an explanation of why he was still citing the paper when he knew that it was flawed.

Jones now had the choice of taking his punishment and correcting the record or standing and fighting a battle that he could never win. As he told some of his colleagues, he was determined to fight:

> I won't be replying to either of the emails below, nor to any of the accusations on the *Climate Audit* website. I've sent them on to someone here at UEA to see if we should be discussing anything with our legal staff...I do now wish I'd never sent them the data after their FOIA request![75]

Having failed to get a response from Wang, Keenan carried through on his threat to make a formal complaint to Wang's employer, the University of Albany. The story of how the complaint was dealt with is remarkable, not only in itself, but also because of the way it presaged what was to follow.[76] The university initially set up a panel of inquiry, the job of which was to determine if there was a case to answer. By early 2008, a report had been issued, its authors concluding unanimously that Wang had not had the data he claimed to have. At first sight, this might have looked as if the formal investigation that followed would therefore be a foregone conclusion – after all, if Wang did not have the data then he must have fabricated his conclusions. However, a few months later, Keenan was informed that the investigation committee had concluded that there was no evidence of fraud.

This was an extraordinary claim, flying in the face of everything that was known about the case. And if there was any doubt over the nature of the investigation, this was fully dispelled by Keenan's subsequent revelation that the investigation panel had not even interviewed him as part of their work, a clear breach of their own rules. Perhaps even more remarkably, the panel then invited Keenan to comment on the findings, while refusing to actually let him see their report, apparently on the grounds that he had not been interviewed during the investigation. There was little Keenan could do, although he made a protracted effort to force regulators to become involved, issuing complaints to the Attorney General of New York and the Department of Energy. As with his complaint to the university, however, each one was rejected without explanation.

THE LAST THROW OF THE DICE

With UEA's first brushes with the reality of the FOI Act having been so disastrous, and with David Holland's requests causing further problems, the CRU scientists continued to explore ways of avoiding compliance with the legislation. In the second half of 2008, Jones updated some of his colleagues about what had been going on and his message contains the remarkable revelation that the Information Commissioner's office had been helping them to avoid FOI requests.

> Keith/Tim still getting FOI requests as well as [the Met Office and Reading University]. All our FOI officers have been in discussions and are now using the same exceptions not to respond – advice they got from the Information Commissioner.[77]

The culture at UEA appears to have spun quite out of control, with the breaking of FOI legislation becoming a feature of everyday life in the university. In the same email Jones said that he had 'deleted loads of emails', a potentially dangerous thing to do given the terms of the FOI legislation. What is worse, there is evidence that the university's FOI officer David Palmer had been persuaded to help the scientists avoid compliance with the law and that even the university's vice-chancellor was party to this effort. In an email to Santer and Wigley at the end of the year, Jones indicates how Palmer, the man charged with ensuring the application of FOI legislation at UEA, had been turned.

> When the FOI requests began here, the FOI person said we had to abide by the requests. It took a couple of half hour sessions – one at a screen, to convince them otherwise showing them what [*Climate Audit*] was all about. Once they became aware of the types of people we were dealing with, everyone at UEA (in the registry and in the Environmental Sciences school – the head of school and a few others) became very supportive. I've got to know the FOI person quite well and the Chief Librarian – who deals with appeals. The [vice-chancellor] is also aware of what is going on – at least for one of the requests,* but probably doesn't know the number we're dealing with. We are in double figures.[78]

There was no let-up in the pressure. In 2009, McIntyre tried once again to get the raw data for the CRUTEM temperature series. His first port of call was not UEA but the Met Office which, as we have seen, combines CRU's land temperature data with ocean temperature records to calculate the HADCRUT global temperature index. The Met Office's response was that the data was held under confidentiality agreements and could not

---

*The vice-chancellor at that time was Bill MacMillan. He stood down through ill health in 2009.

therefore be disclosed. They said, however, that these agreements had been lost and that therefore they could not say which countries were affected. Rebuffed, McIntyre decided to try CRU again, but this time he had a possible ace up his sleeve: he had discovered that Jones had sent the CRUTEM data to Peter Webster, a climatologist in America. If Webster could get the data, there was surely no reason that McIntyre could not have it too.

McIntyre's new request presented CRU with a serious problem: they would now have to find a reason why McIntyre could not have the data while Webster could. Several different approaches were tried. The first refusal notice issued by the university claimed that the data was held under agreements that prevented onward transmission to non-academics. This was a remarkable story, given that the Met Office had previously told McIntyre that the agreements had been lost. However, their excuse was quickly shown to be just that, when Ross McKitrick and a number of other academic readers of *Climate Audit* sent their own requests for the same data. These applicants were told that the story about non-academics had been a mistake and that in fact the data could not be forwarded to anyone – apparently when UEA had sent the data to Webster this had been another mistake.

The response was obvious, and a decision was made to ask for copies of the agreements themselves. *Climate Audit* readers started sending requests, asking for five countries each. Their smokescreen having been seen through, CRU's hand was forced and they responded with a notice explaining their problem. This was a remarkable document, stating that the university had disposed of the original raw data during the 1980s due to data storage issues arising from a lack of resources. As for the agreements, requesters were directed to a webpage containing just a handful of documents, although CRU claimed that there were others that had, like the data, been lost during office moves during the 1980s. If this pretext was not flimsy enough, few, if any of the agreements that UEA were able to produce actually prevented further dissemination of the data anyway, most simply requiring the source to be cited if the data was used.[79]

It was a sorry state of affairs – once again, attempts to block FOI requests had brought UEA into disrepute, at least among the small band of sceptics. However, it was not long before the news was to reach a wider audience.

POLAR URALS AND YAMAL

As interest in the request for Webster's data faded, the sceptics' attention shifted back to paleoclimate. The saga of the Polar Urals and Yamal tree ring series stretched back to the earliest days of McIntyre's involvement in the field and has been told in detail elsewhere.[80] The hockey-stick-shaped Polar Urals series was an almost ubiquitous ingredient in early tree-ring-based temperature reconstructions. However, in 1999 the series had been updated and was found to no longer have the same shape – the update suggested strong medieval warmth. Thereafter, paleoclimatologists had taken to using another hockey-stick-shaped series from the same region – Yamal – and in fact Polar Urals had almost never been used again, despite the fact that it was more reliable – 'better replicated' in the jargon.

The Yamal data had been collected by two Russian scientists named Hantemirov and Shiyatov, but had become important through its development into a series by CRU's Keith Briffa. McIntyre had been trying to get hold of the raw measurement data behind Yamal for many years. He had been able to get one version of the data from the Russians but, while this was interesting, it was not exactly what he needed because the Russians' figures were not necessarily the same as Briffa's. What the Russian data did reveal, however, was that the number of tree ring cores behind the twentieth century uptick at the end the Yamal hockey stick was astonishingly low – it was in single figures by the end of the twentieth century. This was very strange because this should have been the point at which it was *easiest* to get data. However, without getting his hands on the actual data used, it was impossible for McIntyre to show that Briffa's version of Yamal suffered from a similar paucity of data, and Briffa was not inclined to help; McIntyre's requests to CRU had always been refused and the journals in which Briffa had published also refused to force disclosure. Eventually, however, Briffa slipped up.

Tree ring chronologies are normally put together using data collected from a relatively small geographic area. However, in 2006 Briffa started to look at what happened when data from a much larger area was used. His research was to focus on three of these so-called 'regional chronologies', all of which were located in northern Europe: Fennoscandia, Avam–Taimyr, and Urals. When the calculations were complete, Briffa will have been concerned to see that the Urals chronology, which incorporated the Yamal series, no longer had a hockey-stick shape. This was doubly problematic, since not only did it contradict the idea of unprecedented warmth that his colleagues in the IPCC were so keen on, but it also underlined the magnitude of the so-called 'divergence problem'.

The divergence problem is a critical issue for paleoclimatologists. In essence, if the modern warming is unprecedented, tree-rings should be reaching unprecedented sizes and densities. However, some tree rings are are not responding to recent temperature rises in the expected way and nobody has been able to understand why. This contradictary evidence raises the disturbing question of whether tree rings are actually responding to temperature at all; if they are not then they cannot be used to reconstruct temperatures of the past and the paleoclimate temperature reconstructions would have to be treated as unreliable. Climatologists were understandably keen to argue that only a small number of tree ring series were affected, and so the Urals regional chronology would have been a blow to Briffa. What happened next suggests that Briffa felt that what he had found was so unacceptable that it had to be suppressed.

The results of Briffa's work were eventually published in the prestigious *Philosophical Transactions of the Royal Society B* in 2008,[81] in a special edition devoted to the northern forests. However, while the paper referred to three regional chronologies, Briffa had decided not to use the Urals chronology. Instead he presented what was referred to as a Yamal regional chronology, but was in fact just the Yamal series on its own. This sleight of hand enabled Briffa and his co-authors to present a headline-grabbing set of conclusions, none of which would have been supportable if the Urals chronology had been used:

> These data show no evidence of [the divergence problem]...as has been found at other high-latitude Northern Hemisphere locations...

> [We] quantify the time-dependent relationship between growth trends of the long chronologies as a group. This provides strong evidence that the extent of recent widespread warming across northwest Eurasia, with respect to 100- to 200-year trends, is unprecedented in the last 2000 years.[81]

Briffa may have felt that the divergence problem in the Urals had been swept under the carpet but he was soon to be disabused of this idea. Shortly after the paper's publication, McIntyre realised that the

Royal Society's publishing arm had what appeared to be a clear and robust policy on data availability, which required authors to archive their data at the time of publication. When he took the matter up with the Royal Society, he was gratified to be told that they took such matters very seriously and even received an apology for the failure to notice Briffa's non-compliance.

Later that year Briffa was forced agree to disclose the data, although it was only a year later, in September 2009, that he finally complied. When the figures were published the extraordinary lack of data underlying the blade of the Yamal hockey stick caused a minor sensation. In fact the high point at the end of the graph was shown to have been based on only *four* trees, and only one of these had the hockey stick shape. McIntyre dubbed it 'the most influential tree in the world'.[82]

However, it was not only the lack of data that had attracted McIntyre's attention. Through some laborious detective work he was able to determine that the Avam–Taimyr regional chronology incorporated some data series developed by the Swiss researcher, Fritz Schweingruber, but this then raised the question of why Briffa had not used the Schweingruber series that had been collected in the vicinity of Yamal – there was one in particular called Khadtya River that was close by, and with 34 cores represented a much more reliable basis for a temperature reconstruction than Yamal.

By way of an experiment, McIntyre decided to perform a sensitivity analysis on the Yamal series, replacing the 12 cores that were used in its twentieth century section with the 34 cores from Khadtya River. The results were once again devastating, with the blade of the Yamal hockey stick replaced by a sharp downturn.

There was an almost immediate response from the Hockey Team. *RealClimate*, a blog run by Mann and some other members of the Hockey Team in order to defend the mainstream position on climate change, published a fierce attack, ridiculing McIntyre's use of Khadtya River.

> McIntyre has based his 'critique' on a test conducted by randomly adding in one set of data from another location in Yamal that he found on the internet. People have written theses about how to construct tree ring chronologies in order to avoid end-member effects and preserve as much of the climate signal as possible. Curiously no-one has ever suggested simply grabbing one set of data, deleting the trees you have a political objection to and replacing them with another set that you found lying around on the web.[83]

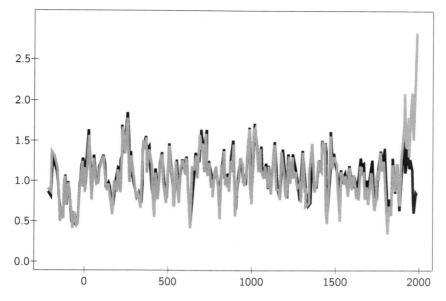

FIGURE 2.1: The Khadtya River experiment

Briffa's Yamal series is shown in grey and McIntyre's experiment incorporating the Khadtya River series is in black. The twentieth century uptick in the Briffa series is replaced with a sharp downturn.

However, when Briffa himself responded a few weeks later, he took a slightly different line to his colleagues at *RealClimate*, acknowledging that Khadtya River met the criteria for inclusion in the a regional chronology in the area, but claiming that he and his co-authors had not considered it at the time.

> Judged according to [our normal] criterion it is entirely appropriate to include the data from the [Khadtya River] site...when constructing a regional chronology for the area. However, we simply did not consider these data at the time, focussing only on the data used in the companion study by Hantemirov and Shiyatov and supplied to us by them.[84]

Briffa's statement was simply untrue; not only had he considered the Khadtya River series, but he had included it in the Urals regional chronology that he had refused to publish in his paper. This was not the end of the deception either. Briffa's rebuttal also included what he said was a revised Yamal chronology, produced 'by making use of all the data to hand'. This, he showed, gave broadly the same result as the one he had published in his paper. However, Briffa was again being economical with the truth. His revised chronology, which allegedly included 'all of the data to hand', actually included just a small fraction of the data in the Urals chronology. Of course, nobody outside Briffa's circle was aware of the existence of this larger regional chronology, and Briffa made no mention of it in his article. For the time being, therefore, Briffa was able to get away with his deceit. Eventually, however, it would be his undoing.

Briffa had held the line, but it was clear that Yamal was rapidly turning into yet another embarrassment for CRU. To Jones, Briffa and Osborn it must have seemed as if there was no end to the humiliations they were suffering at the hands of the sceptics. Whichever way they turned they seemed to be beset by demands for the release of their data and the publication of their code. No matter how they responded there were always more questions, more demands: it must have seemed as if their stranglehold on the climate debate was in danger of being wrenched from their grasp. But even they can have had no idea just how close the end was. Just days later, the world of the CRU scientists was to change forever.

CHAPTER 3

CLIMATEGATE

At around 07:20 EST, on 17 November 2009, somebody hacked into the servers hosting *RealClimate*, the semi-official blog that has been at the centre of efforts to defend the IPCC consensus and to attack sceptic critiques of mainstream climatology. Having gained access, the hacker uploaded a large zip file containing over 1000 emails taken from the servers of CRU. Ten years of correspondence between the members of the Hockey Team had somehow been released into the public domain, along with folders containing a wealth of documents, data and code that had long been sought by climate change sceptics. After uploading the zip file, the hacker composed a draft post for publication on the *RealClimate* blog:

> We feel that climate science is, in the current situation, too important to be kept under wraps. We hereby release a random selection of correspondence, code, and documents. Hopefully it will give some insight into the science and the people behind it.
>
> This is a limited time offer, download now.

This was followed by a link to the zip file and a series of extracts from the emails:

```
...
1225026120.txt * CRU's truncated temperature curve
1059664704.txt * Mann: dirty laundry
1062189235.txt * Osborn: concerns with MBH uncertainty
0926947295.txt * IPCC scenarios not supposed to be realistic
...
```

A day earlier, McIntyre had posted up a new blog article at *Climate Audit*, in which he reported on a presentation given to the Canadian Society for Petroleum Geologists by Brian Luckman, a paleoclimatologist from the University of Western Ontario.[85] The subject of the talk had been the many different factors that can affect tree growth, and how difficult it is to use tree rings for temperature reconstructions without controlling for all these different effects – he had described this fraught process as 'a black art'.

McIntyre was clearly amused by this characterisation, which rather seemed to confirm many of his criticisms of the reliability of tree rings as the basis for temperature reconstructions. To underline the point he added a cartoon at the end of the post, showing two scientists looking at an algebra-filled blackboard, in the centre of which were the words, 'Then a miracle occurs'. One scientist was saying to the other: 'I think you should be more explicit here in step two'.

It was a small joke at the expense of the paleoclimatologists and as such it drew little or no comment. For most of the following 24 hours, *Climate Audit* readers engaged in a vigorous discussion of the intricacies of tree ring growth and temperature reconstructions, just as they did when commenting on any of McIntyre's other postings, and it is perhaps not surprising that nobody seems to have noticed the single-line comment that appeared at 07:24 EST:

> A miracle just happened.[86]

The commenter gave their user name as 'RC', which should have alerted readers to something unusual: to the *Climate Audit* readership, the acronym RC is a clear reference to *RealClimate*, and it is extremely unusual for any of the authors there to comment at McIntyre's site. But there was another clue that this was something unusual: readers who investigated the comment further might have noticed that instead of giving a normal website address, RC had created a link to the zip file that had been uploaded to *RealClimate* just minutes earlier. And if linking to a file rather than a blog was not enough, the name of the file – 'FOIA.zip' – would also have set alarm bells ringing because in normal usage the acronym refers to the US Freedom of Information Act. Despite all these signals, however, or perhaps because the link formed part of the commenter's name rather than appearing in the body of the comment text, it appears that nobody paid the comment any attention. The *Climate Audit* readers continued discussing tree rings for the rest of the day, unaware of the gift that had been placed at their feet.

PLAN B

Even if the comment had been noticed, it is likely that the hacker's plans would have been thwarted anyway. The strategy was perhaps slightly too elaborate; there were too many hostages to fortune. Before the draft posting for *RealClimate* could be published, the hacker's presence on the site was noticed. Gavin Schmidt, the climate modeller who is one of the mainstays of the site, has described what happened:

> I tried to log in to *RealClimate*, but for some reason my login did not work. Neither did the admin login. I logged in to the back-end... only to be inexplicably logged out again. I did it again. No dice. I then called the hosting company and told them to take us offline until I could see what was going on. When I did get control back from the hacker (and hacker it was), there was a large uploaded file on our server, and a draft post ready to go announcing the theft of the CRU emails. And so it began. [87]

Realising that the posting was unauthorised, Schmidt quickly blocked it from publication before it could appear. It looked as though the hacker's carefully conceived plan would come to nothing.

That evening, however, a back-up plan was put into action. While it clearly would have been amusing to have sceptics discover the zip file on *RealClimate* itself, in reality there was no need to place it on the biggest mainstream climate blog. That evening, the hacker paid visits to three of the most important sceptic blogs in turn – *Watts Up With That?*, *Climate Skeptic* and *The Air Vent*. [88] At each one he posted the same comment, the wording of which was identical to what he had tried and failed to post at *RealClimate*. However, with the zip file having been removed from *RealClimate*, it had clearly been necessary to find another home for all the CRU information, and the new comments now pointed to a copy of the data located on a server in Russia.

Even though this new approach clearly stood a much better chance of success, things still did not run entirely smoothly. Like the *Climate Audit* readers, Warren Meyer, the host of the *Climate Skeptic* blog, failed to notice the comment at all, while Jeff Id, the pseudonymous author of *The Air Vent*, had gone hunting a few days previously. Almost unbelievably, none of the readers of either site appear to have realised the importance of what had been posted. In fact, even when Id returned from his trip, he still seems to have missed the all-important comment, posting up two articles on unrelated matters once he got back to his desk.

Fortunately, at *Watts Up With That?*, the story was different, if far from straightforward. The site, run by former TV meteorologist Anthony Watts, is the most popular climate blog in the world. By coincidence, Watts was also away from his desk at the time that the hacker left his comment, attending a conference of global warming sceptics in Europe.

However, because of its prominence in the climate debate, *Watts Up With That?* receives hundreds of comments on every posting and the conversations are often protracted and heated. Because of this, the site has a team of moderators who try to ensure good behaviour. When the hacker posted his comment about the zip file, therefore, it was immediately noticed by the head of Watts' moderation team, Charles Rotter, known to readers simply at 'Charles the Moderator'. Rotter quickly realised that the comment could be important but he was also alarmed by the fact that it contained a link to a Russian server. Russian websites usually spell trouble for bloggers, often leading readers to inappropriate or dangerous material such as pornography and phishing sites. With this risk in mind, Rotter quickly put the comment in quarantine and flagged it for the attention of the other moderators so that nobody would inadvertently release it again. Then, somewhat nervously, he downloaded the file, scanned it for viruses and started to read.

It didn't take long for Rotter to realise the importance of what the zip file contained and within minutes he was emailing Anthony Watts, who was about to make a presentation at the European Parliament in Brussels. Once Rotter had explained the zip file to him and reassured him that it was safe to view the contents, Watts quickly downloaded everything to see for himself. It was clear that they were sitting on a huge story but the question of how to proceed was a difficult one. For a start, the legal implications of publishing leaked or hacked material were not to be taken lightly and Watts desperately wanted to take advice from a lawyer before taking any futher action. But there was also the possibility that the zip file was some sort of an elaborate fraud – perhaps a trap to discredit *Watts Up With That?* in some way. These concerns were multiplied by the fact that in just a few hours time Watts was due to go through the elaborate security procedures at the European Parliament. Why was this all happening now, he wondered. Opting for caution, he downloaded a program that would enable him to wipe all trace of the zip file from his laptop, before proceeding with a great deal of trepidation to the parliament building. Breaking the story to the public could wait until he he was safely back home.

The next 48 hours were a highly stressful experience for Watts:

> I surely must have looked like a guilty person due to my nervousness, but I was relieved to find myself able to get through security (which was a long process), and I attended the conference. The conference itself was a huge burden on me, because here I was,

holding a terrible secret in a room full of the most prominent skeptics in the world, and I couldn't say a word...

I finished my presentation, went back to my hotel, and stayed there. I took in no sights, I didn't go out to eat, and took my meals in my room. All I wanted to do was get back to the USA.[89]

While Watts finished his business in Brussels, there was plenty for Rotter to do back in San Francisco. He and Watts had agreed that they should delete the comment from the website so as to be absolutely certain that it wouldn't be inadvertently released onto the web by one of the other moderators. Having done so, the next step was to try to verify the authenticity of the contents. In order to do this, Rotter brought two key people into the secret of the zip file's existence. The first of these was his roommate, Steven Mosher. Mosher is well known around the climate blogs, commenting regularly at the sites on both sides of the debate, particularly the science-heavy ones such as *Climate Audit* and *The Blackboard*. Mosher was handed a CD with a copy of the emails and was given the task of contacting Steve McIntyre: as an addressee in many of the messages, McIntyre would be able to help them assess if the contents were genuine. However, before he would hand the disk over, Rotter made Mosher promise not to send a copy to anyone, including McIntyre. His only concession was to say that McIntyre could have a copy by mail, but only on condition that it arrived after Watts' return from Europe. Both Mosher and McIntyre were to be kept in the dark about the file's online location.

Mosher's first telephone call to McIntyre, late in the evening of 17 November, seems to have been profoundly shocking for the Canadian: only weeks later he could barely recall the details of the conversation. That evening and for much of the next day, Mosher read message after message, with McIntyre checking the details against his original email correspondence. Much of this had already been published on *Climate Audit*, but many of the messages remained private, so their appearance in the zip file seemed to suggest that the contents might well be genuine. The excitement was almost too much, and Mosher begged Rotter for permission to go public with what they had, but he was met with a flat refusal. As Rotter put it, if it was as important as it appeared to be then two or three days would not make any difference.

THE ALARM

Shortly after the attempt to release the zip file through *RealClimate*, word was sent to UEA that their systems had been compromised. The news quickly spread around the university as an email warning was circulated around all the staff in the School of Environmental Sciences:

> A large volume of files and emails from computers [in the UEA School of Environment] and CRU have been posted onto a web site, apparently by climate change sceptics.

> We are seeking advice from [the IT department] as a matter of urgency...

One of the recipients of the email was Paul Dennis, the head of UEA's Stable Isotope and Noble Gas Geochemistry Laboratories. As we have seen, CRU is a unit within UEA's School of Environmental Sciences, and Dennis was therefore a colleague of Jones, Briffa and Osborn, although he was not on CRU's staff. Dennis is a geochemist rather than a climatologist, although his work often involved him in paleoclimatology and he had published many papers in the area. He was therefore fully aware of the controversies that swirled around CRU. Nevertheless he was not an active participant in the climate wars and even enjoyed cordial relations with the sceptic community: he had been an intermittent commenter at *Climate Audit* over the years and just a week earlier had emailed some of his work to Jeff Id at *The Air Vent* blog.

The email advising of the disclosure of the CRU emails was as much of a surprise to Dennis as to everyone else. Intrigued, he forwarded the message to McIntyre to let him know what was happening. McIntyre of course knew what was going on already, but had to content himself with assuring Dennis that he would get in touch if he heard anything.[90] If McIntyre and Mosher still had any remaining doubts about the authenticity of the emails, Dennis's message was to lay them to rest. There could now be little doubt that CRU had been hacked and the zip file was genuine. It was just a matter of time before the balloon went up.

Having seen the UEA notice, Mosher, Rotter and McIntyre knew that there was a copy of the zip file somewhere out on the web. In essence then, the zip file was already a public document: if they could locate it they would no longer have to wait for Anthony Watts to return from Europe but could go public immediately. Rotter therefore put the URL of the Russian server into a search engine and moments later turned up the hacker's message at *The Air Vent*, complete with its link to the Russian server – the zip file was indeed out in the wild and any confidentiality agreement with Anthony Watts was now superseded. Soon afterwards, they started to spread the word.

The first person to find out was Jeff Id at *The Air Vent*. Id, like Watts, was very concerned about having the link to the zip file on his site and quarantined the comment while he worked out what to do. Meanwhile, however, Mosher had also announced his discovery at another prominent sceptic blog, Lucia Liljegren's *Blackboard*.

> Lucia,
>
> Found this on Jeff Id's site.
>
> http://noconsensus.wordpress.com/2009/11/13/open-letter/
>
> It contains over 1000 mails. IF TRUE.
>
> 1 mail from you and the correspondence that follows.
>
> And, you get to see somebody with the name of Phil Jones say that he would rather destroy the CRU data than release it to McIntyre.
>
> And lots lots more. Including how to obstruct or evade FOIA requests. . . and guess who funded the collection of cores at Yamal. . . and transferred money into a personal account in Russia.
>
> And you get to see what they really say behind the curtain. You get to see how they 'shape' the news, how they struggled between telling the truth and making policy makers happy.
>
> You get to see what they say about Idso and Pat Michaels, you get to read how they want to take us out into a dark alley. It's stunning, all very stunning. You get to watch somebody named Phil Jones say that John Daly's death is good news. . . or words to that effect.
>
> I don't know that it's real.
>
> But the CRU code looks real.[91]

Shortly afterwards, Watts posted his own take on the story from Dulles airport in Washington DC, where he was awaiting a flight back to California. Meanwhile, Mosher headed over to *Climate Audit*, where he started posting excerpts from the emails for the benefit of the readers. The word spread like wildfire and the blog was quickly so overwhelmed by visitors that the system was unable to cope. Within minutes even McIntyre was unable to gain access. After many attempts he finally managed to connect with the server and post a one-word comment: 'Unbelievable'. Then he went to play squash.

It was the afternoon of 19 November 2009.

CHAPTER 4

THE STORM

THE STORM BREAKS

It must have been clear to everyone at *RealClimate* that it was only a matter of time before the zip file's existence became public knowledge and the contents were disseminated around the world. Gavin Schmidt certainly seems to have been keeping a close eye on all the sceptic blogs, looking for the first word of the zip file to appear. Within half an hour of the first comment appearing at Lucia's Liljegren's *Blackboard* site, he had intervened to try to stop the floodgates opening, writing an email to Liljegren as follows:

> Subject: A word to the wise
>
> Lucia, As I am certain you are aware, hacking into private emails is very illegal. If legitimate, your scoop was therefore almost certainly obtained illegally (since how would you get [a thousand] emails otherwise). I don't see any link on Jeff Id's site, and so I'm not sure where Mosher got this from, but you and he might end up being questioned as part of any investigation that might end up happening. I don't think that bloggers are shielded under any press shield laws and so, if I were you, I would not post any content, nor allow anyone else to do so. Just my twopenny's worth. [92]

Liljegren had realised the possible implications of having allegations from the emails on her site and at first had 'unapproved' several comments and closed down the thread to further contributions while she worked out how to proceed. But it quickly became clear that the emails would soon be all over the internet and shortly afterwards she posted up the first blog article on the subject. If Schmidt was making an attempt to cut the leak off at source, it failed. With readers already downloading the zip file as fast as the Russian server would allow, it was already too late to prevent the situation spiralling out of control. Even when it was discovered that the file had been removed from the Russian server, there was little concern. As Liljegren put it, 'This genie ain't going back in the bottle'.

The fact that CRU had issued a warning to their staff seemed to show that the contents of the zip file were genuine, but the sceptic blogosphere initially adopted a position of extreme caution, noting that it was quite possible that some of the messages might have been tampered with in some way. The possibility of a trick was topmost in their minds for several hours, and the concern was heightened further when one commenter at *Climate Audit* noted that the file creation dates on some of the data files appeared to have been altered so that they were all the same.[93] Others pointed out, however, that this was likely to be an artifact of the way in which the data had been collected and the worries over the authenticity soon passed – the level of detail in the information released was simply overwhelming. Nobody could possibly have faked it.

The allegations were already starting to flow. The message left by the hacker at the sceptic blogs had pointed to several important issues, but every few seconds something new would be posted on the comments threads at the sceptic blogs – tampering with data, perverting the peer review process, deleting information subject to Freedom of Information requests – the list was apparently unending. For sceptics it was as though Christmas had come early. Here, suddenly, was confirmation of almost all of the criticisms they had been making for the last ten years or more. Inevitably the term 'Climategate' was coined for the scandal and, inevitably, it stuck.

By lunchtime on 20 November, Ian Wishart, a journalist from New Zealand, had managed to make contact with Phil Jones and get a comment on what was happening.* The conversation appears to have been brief, but Jones did give a few details of what had happened inside CRU once the news of the disclosures reached them:

> It was a hacker. We were aware of this about three or four days ago – that someone had hacked into our system and taken and copied loads of data files and emails.[95]

CRU staff may have been aware of the hacking for several days, but Jones said that he had only just discovered that the zip file was publicly available. He also revealed a little of what was going on at UEA at the time. The School of Environment appeared to be in chaos, with passwords having all been changed and systems locked down. Jones and his colleagues were apparently even unable to access their email accounts.

---

*Fred Pearce of the *Guardian* also spoke to Jones at around this time (see *The Climate Files*,[94] p. 53), He was told that the view within CRU was that the disclosure of the emails had something to do with McIntyre's efforts to get hold of the Yamal tree ring data.

Wishart had kept a close eye on the blogs and knew that one email in particular was getting a huge amount of attention – indeed it came to dominate media coverage of the Climategate affair in the weeks and months afterwards. This was an email from Phil Jones to the authors of the Hockey Stick paper – Michael Mann, Ray Bradley and Malcolm Hughes:

> I've just completed Mike's Nature trick of adding in the real temps to each series for the last 20 years (i.e. from 1981 onwards) and from 1961 for Keith's to hide the decline.[96]

Talk of tricks and hiding things was clearly going to raise the worst suspicions, but at first there was considerable confusion over what particular decline it was that was being hidden. Indeed, in the initial mayhem following the release of the emails, many people erroneously assumed that Jones was referring to a decline in global temperatures – in other words that the planet was actually cooling.* In fact, the problem was much more subtle, and related to a graph of tree-ring measurements. The implications were, however, still extremely serious.

As the first journalist to speak to Jones in the wake of Climategate, Wishart did not want to miss the opportunity to get the CRU man's side of this story and he asked for an explanation of what he had meant by 'hide the decline'. Taken by surprise, Jones prevaricated, telling Wishart that he had forgotten the precise meaning:

> That was an email from ten years ago. Can you remember the exact context of what you wrote ten years ago?[95]

However, he elaborated somewhat, explaining that he had been talking about a temperature reconstruction where the underlying data series stopped short of the modern times:

> ...it's just about how you add on the last few years, because when you get proxy data you sample things like tree rings and ice cores, and they don't always have the last few years. So one way is to add on the instrumental data for the last few years.[95]

To a layman, this might at first sight have appeared to be a reasonable, if somewhat garbled, explanation, but a moment's thought would

---

*The most prominent person to repeat this mistaken story was former US vice-presidential candidate, Sarah Palin.

have raised some important question marks. What exactly was declining? Why would you want to add instrumental data onto a proxy series anyway? Surely you should show the two series separately? Why would you want to hide something? And from whom? When examined closely, Jones' explanation raised almost as many questions as it answered.

HOLDING THE LINE

Sceptics' discussions of the trick to hide the decline had clearly been noticed by the some of the movers and shakers in the climate establishment and by the early afternoon of 20 November it was possible to discern the beginnings of a public relations programme in defence of the CRU scientists. Within 20 minutes of each other, two articles appeared – the first responses to the affair from the sceptics' opponents. The first was a blog posting at *RealClimate*, which appears to have been formulated by the whole *RealClimate* team.[97] Gavin Schmidt and his colleagues, among them Michael Mann, had had three days to consider their response and the article gives the impression of having been carefully thought through. There were three main strands to the article. Firstly, emphasis was placed on the illegality of the hack:

> As people are also no doubt aware the breaking into of computers and releasing private information is illegal, and regardless of how they were obtained, posting private correspondence without permission is unethical.[97]

Whether the release of the emails from CRU counted as hacking or was illegal is a moot point. If the person who obtained the emails was an insider – a whistleblower, in other words – then their actions may well have been permissible under UK law. That said, accessing the *RealClimate* servers may well have been another story. Then again, it is also unlikely that the emails were in fact private since the vast majority would have been subject to disclosure under UK FOI legisation.

Having opened the defence by trying to present the CRU scientists as the wronged party rather than by addressing what was in the emails, the *RealClimate* team moved on to discuss what they did *not* contain.

> More interesting is what is not contained in the emails. There is no evidence of any worldwide conspiracy, no mention of George Soros nefariously funding climate research, no grand plan to 'get rid of the [medieval warm period]', no admission that global warming

is a hoax, no evidence of the falsifying of data, and no 'marching orders' from our socialist/communist/vegetarian overlords. The truly paranoid will put this down to the hackers also being in on the plot though. [97]

Instead, it was suggested, the emails revealed scientists 'engaging in "robust" discussions' and 'expressing frustration at the misrepresentation of their work'. [97]

Although the tone of the post was slightly overwrought, it was clear even at this early stage of the investigation into the emails that there was a measure of truth in the explanations given by the *RealClimate* team – the CRU scientists did indeed have a sincere belief in the truth of the manmade global warming hypothesis; there was no evidence of a grand conspiracy to defraud the public. However, the argument that what the emails revealed was merely loose talk caused by scientists' frustration at their critics was far from the truth. As we will see, there was abundant evidence of a wrongdoing that was more subtle in nature than a grand conspiracy but nearly as damaging – evidence that scientists were willing to twist the facts to support their case and to crush anyone who disputed what they said. We will also see that the emails contained hints that there *was* a wish among some of the scientists involved to 'get rid of the Medieval Warm Period'.

There was another narrative that appeared in this article for the first time but was to be heard repeatedly in the coming months. This was the idea that sceptics were quoting the emails in isolation, the insinuation being that in their proper context the conduct of the scientists would appear perfectly innocent. Despite this, however, the mitigating context never appeared. Indeed, as often as supporters of the global warming cause repeated their mantra that the emails were being quoted out of context, sceptics would repeat theirs: that the context only made things look worse.

Although they could try to set a narrative context to guide the future debate, the *RealClimate* team were clearly going to have to provide an answer to at least some of the more serious allegations – these were being made in their own comments threads and across the web and demanded a response. With so much of the noise on the internet focusing on the trick to hide the decline in the tree-ring measurements, Gavin Schmidt and his *RealClimate* colleagues had little choice but to address this question before it gained too much headway in the media. The story they formulated was that Jones' words were being taken out of context and that his actions were entirely innocent.

> ...the 'trick' is just to plot the instrumental records along with reconstruction so that the context of the recent warming is clear. Scientists often use the term 'trick' to refer to a 'a good way to deal with a problem', rather than something that is 'secret', and so there is nothing problematic in this at all. As for the 'decline', it is well known that Keith Briffa's [MXD] proxy diverges from the temperature records after 1960 (this is more commonly known as the 'divergence problem'...and has been discussed in the literature since Briffa *et al.* in Nature in 1998...Those authors have always recommend not using the post-1960 part of their reconstruction, and so while 'hiding' is probably a poor choice of words (since it is 'hidden' in plain sight), not using the data in the plot is completely appropriate, as is further research to understand why this happens.[97]

This was clearly not going to be the end of the affair, and we will examine the story of the trick to hide the decline and the *RealClimate* explanation of it a little later.* In the meantime, however, there was more to the media defence of CRU. Just minutes after the *RealClimate* post appeared, a second defence of Jones and his colleagues was published.[98] The article, by Leo Hickman and James Randerson, appeared on the *Guardian*'s website just twenty minutes after Gavin Schmidt's effort at *RealClimate*, and it is instructive to compare the two postings.†

Hickman and Randerson began by setting out the background to the Climategate story and then went on to quote a UEA spokesman who had confirmed the release of material. However, the spokesman also raised the question of whether everything circulating on the internet was genuine. Then, just as Schmidt had done, Hickman and Randerson made much of the emails' disclosure being illegal, a point they reinforced with a typically combative quote from Michael Mann:

> I'm not going to comment on the content of illegally obtained emails. However, I will say this: both their theft and, I believe, any reproduction of the emails that were obtained on public websites,

---

*See p. 83.
†There is no time of posting on the *RealClimate* article, but the first comment appeared at 12:55 EST, which is 17:55 GMT. The *Guardian* article appeared at 18:15 GMT, just 20 minutes later.

etc, constitutes serious criminal activity. I'm hoping the perpetrators and their facilitators will be tracked down and prosecuted to the fullest extent the law allows.[98]

Mann and Schmidt are both *RealClimate* authors and so they would undoubtedly have discussed how best to respond to the email disclosures – it is thus no surprise to see them making similar cases. However, the *Guardian* article contained another echo of Schmidt's article that throws some light on what was happening behind the scenes. This related to the *Guardian*'s explanation of the trick to hide the decline:

> This sentence [about the trick to hide the decline], in particular, has been leapt upon by sceptics as evidence of manipulating data, but the credibility of the email has not been verified. The scientists who allegedly sent it declined to comment on the email.[98]

In fact, as we have seen, Jones had in effect admitted the veracity of the email in his interview with Wishart. However, in the absence of a comment from Jones, Hickman and Randerson had approached Bob Ward, the public relations director of the Grantham Institute at the London School of Economics. The Grantham Institute is an environmental research organisation, funded by a wealthy financier. Ward, its public face, is well known among those who follow the climate debate, where he occupies the position of the scourge of the sceptics, appearing regularly in the media attacking anyone who questions global warming orthodoxy. Ward's comment on the trick to hide the decline was as follows:

> It does look incriminating on the surface, but there are lots of single sentences that taken out of context can appear incriminating... You can't tell what they are talking about. Scientists say 'trick' not just to mean deception. They mean it as a clever way of doing something – a short cut can be a trick.[98]

The similarity between the argument put forward by Ward and that put forward by Gavin Schmidt on *RealClimate* is remarkable. However, combined with the fact that the two articles appeared only twenty minutes apart it is hard to escape the conclusion that Hickman, Randerson and Ward were part of a wider coordinated effort to manage media reactions to the Climategate affair.

It was not only *RealClimate*'s sympathisers in the media who were taking notice of the Climategate story. The first articles from the handful of sceptic writers in the mainstream media appeared the same day, with James Delingpole of London's *Daily Telegraph* and Andrew Bolt of the *Herald Sun* in Australia both writing headline-grabbing articles about what was going on in the blogosphere.[99,100] Bolt was almost incredulous at some of what he was reading, wondering if the revelations were simply too good to be true and at first advising his readers to treat them with care. Meanwhile Delingpole went as far as to wonder if Climategate would not turn out to be the 'final nail in the coffin of anthropogenic global warming'. He advised investors to sell shares in renewable energy companies as quickly as they could.

By the following day, the story had spread to America, with the *New York Times*, *Washington Post* and Fox News the first US outlets to report the scandal.[101–103] The *New York Times*' Andy Revkin had been one of the first to be made aware of the story by Mosher, and he was also mentioned in several of the emails, so he had something of an insider's view of what had happened. Revkin's news article was lengthy, with comments from a variety of those involved, including Mann and McIntyre. However, it had few details on the specific allegations and, in a blog posting over the weekend, Revkin said,

> The documents appear to have been acquired illegally and contain all manner of private information and statements that were never intended for the public eye, so they won't be posted here. But a quick sift of skeptics' web sites will point anyone to plenty of sources.[104]

Revkin quickly found himself under attack for refusing to publish any of the emails, a position he apparently adopted on the advice of his employer's lawyers.* It is, however, possible that he was coming under pressure to hold the line and do what he could to minimise the impact of the Climategate story. It is known that at least one member of the *RealClimate* blogging team, Raymond Pierrehumbert, had sent Revkin an email that was critical of the way the *New York Times* was handling the Climategate affair.[105] Although the tone was friendly, Pierrehumbert called

---

*This apparent refusal appeared even more culpable when the paper showed itself to have no qualms about publishing the Wikileaks cables at the end of 2010.

what had happened 'cyberterrorism' and 'a criminal act of vandalism and of harassment', and went on to say that Revkin's lack of indignation was 'disconcerting'.

More significant than the media management going on in outlets sympathetic to the CRU scientists was the interview in the *Daily Telegraph* with Lord Lawson who, as Nigel Lawson, had been Britain's Chancellor of the Exchequer under Margaret Thatcher.[106] Although no longer as much in the public eye as he had been twenty years earlier, Lawson remained active as a legislator and a highly influential political figure. In the interview Lawson revealed a plan to form a new global warming think tank, the purpose of which was to 'act as a check on sweeping environmental reform', or so the article put it. Although there were no more details, at the start of the following week the paper carried another story about the new organisation, which Lawson and his colleagues had decided to call the Global Warming Policy Foundation (GWPF). More interestingly, the article revealed that they had realised the importance of the Climategate disclosures and were calling for a public inquiry into the events at CRU.

> This morning Lord Lawson, who has reinvented himself as a prominent climate change sceptic since leaving front line politics, demanded that the apparent deception be fully investigated.
>
> He claimed that the credibility of the ... world-renowned [CRU] – and British science – were under threat.[107]

The article quoted Lawson as saying:

> They should set up a public inquiry under someone who is totally respected and get to the truth...If there's an explanation for what's going on they can make that explanation.[107]

It is not clear whether the timing was coincidence or not, but within hours of Lawson's call being made public, UEA had responded with a press release of their own, the first official statement on the Climategate affair from the university. In it they announced that they would be empanelling their own inquiry to look into the allegations, although they were giving away few details:

> ... we will ourselves be conducting a review, with external support, into the circumstances surrounding the theft and publication of this information and any issues emerging from it.[108]

The university also made reference to some of the allegations circulating in the blogs, and their arguments are remarkable for their similarity to those made in the *RealClimate* and *Guardian* articles – that some of the emails might not be genuine, that many of them were being taken out of context and that the word 'trick' was actually a reference to 'a clever thing to do'.

While this appeared to add little to what had already been reported in the media, with the university now confirming the veracity at least of many of the emails, the rest of the mainstream media appear to have decided, if apparently somewhat reluctantly, that the story was not going to go away. By the evening of 23 November, nearly a week after the zip file was first posted on the web, the media floodgates had opened. At first it was just a trickle of stories, but the volume grew steadily for many days to come, and it was to be many months before Climategate finally slipped from the headlines.

JONES SPEAKS

Apart from his brief engagement with the media after the disclosures, Jones appears to have decided to keep as low a profile as possible, and none of the journalists reporting on Climategate managed to get an interview. However, with the storm breaking in the media, the university had no choice but to put their side of the story and they issued a lengthy press release featuring statements from university officials and Jones himself.

Jones appeared keen to maintain the idea that he was the wronged party, portraying CRU as having been the victim of a theft and complaining of being bombarded with FOI requests. In a partial concession to his critics, however, he admitted that the emails looked incriminating:

> My colleagues and I accept that some of the published emails do not read well. I regret any upset or confusion caused as a result. Some were clearly written in the heat of the moment, others use colloquialisms frequently used between close colleagues.
>
> We are, and have always been, scrupulous in ensuring that our science publications are robust and honest.[109]

Some favoured media outlets received written statements, and these mostly repeated the 'loose talk' narrative that had been seen in the *Real-Climate* posting and in the UEA statement. For example, in a statement issued to Channel Four News in the UK, Jones explained:

Some of the emails probably had poorly chosen words and were sent in the heat of the moment when I was frustrated. I do regret sending some of them.[110]

The Channel Four interview was also interesting because it showed that Climategate was getting noticed in UK political circles. The station had recorded an interview with the chairman of the House of Commons Science and Technology Committee, Phil Willis, whose words were striking in that they repeated some of the narratives emanating from *RealClimate* and its media sympathisers:

> I think scientists are no different to journalists, to politicians, [or to] schoolteachers, who, when they're communicating with each other, use a language which is certainly looser than if you are writing an official report. Having said that, I think that given the fact we have 24-hour media, that we have emails that can be traced, that it's important in future that they use a language that cannot be misinterpreted.[110]

The report also made clear that UEA were going to try to stand by Jones and that he would remain in position. The university may have hoped that presenting a united front in this way would put some doubt into the minds of the press corps, but the story was spreading out of control regardless. The *Wall Street Journal* said that readers now had 'hundreds of emails that give every appearance of testifying to concerted and coordinated efforts by leading climatologists to fit the data to their conclusions while attempting to silence and discredit their critics'.[111] The *Financial Times* spoke of 'people calling for enormous changes in how the world's economies work discussing ways to keep their data private, manipulate public opinion, and deny dissenters access to the professional literature'.[112]

Among all the media coverage there was a great deal of damage limitation going on. Climatologists, others involved in the IPCC process and those whose livelihoods depended on the existence of the global warming hypothesis had to protect that hypothesis at all costs; the wider scientific community needed to defend the reputation of science itself. Bob Watson, chief scientist at Britain's environment ministry and a former head of the IPCC, was quoted as saying that 'evidence for climate change is irrefutable. The world's leading scientists overwhelmingly agree what we're experiencing is not down to natural variation'.[98] A Greenpeace spokesman said that 'If you looked through any organisation's emails from the last ten years you'd find something that would raise a few eyebrows. Contrary to what the sceptics claim, the Royal Society, the US National Academy of Sciences, NASA and the world's leading atmospheric scientists are not the agents of a clandestine global movement against the truth.'[98]

Few scientists were willing to put their heads above the parapet and criticise their colleagues, a reticence that was understandable in the light of some of the allegations of crushing of dissent that were coming out of the emails. There were a few exceptions, however. In particular, a reaction from the small group of scientists who admit publicly to being sceptics was to be expected, and Roy Spencer, a climatologist from the University of Alabama, was quick to note that the Climategate evidence looked problematic for the mainstream:

> ...there seems to be considerable damning evidence that data have been hidden or destroyed to avoid Freedom of Information Act (FOIA) data requests; data have been manipulated in order to get results that best suit the pro-anthropogenic global warming agenda of the IPCC; e-mails that contain incriminating discussions are being deleted.[113]

But as well as the 'usual suspects', a few less well-known names took the plunge and made public declarations of concern. Eduardo Zorita, a Spanish climatologist based at the Institute of Coastal Research at Geerstacht in Germany, called for Jones, Mann and one of their associates – a climatologist named Stefan Rahmstorf – to be banned from the IPCC process.[114] His call came, Zorita said, 'because the scientific assessments in which they may take part are not credible anymore'. He described the 'machination, conspiracies, and collusion' that were found in some areas of climatology. The scientific debate, he said, had been 'hijacked to advance other agendas'. Zorita's stand was a brave one, and he was quite clear that he expected there to be retribution for his speaking out:

> By writing these lines I will just probably achieve that a few of my future studies will, again, not see the light of publication.[114]

Meanwhile, Demetris Koutsoyiannis, a hydrologist from the University of Athens, said that he was unsurprised by the Climategate revelations, which confirmed many of the conclusions he had reached while struggling to publish a paper that challenged mainstream thinking on climatology.[115]

I must say that what I've been reading in the recently hacked and released confidential files from the CRU...is not a surprise to me. Rather, and sadly, it verifies what I had suspected about some in the climate establishment. [115]

Another scientist risking the possibility of ostracism was Judith Curry, professor of climatology at Georgia Institute of Technology in the USA. Curry is no sceptic, but is something of a free thinker, and in the months before Climategate had written articles for *Climate Audit*, as part of a outreach programme in which she tried to engage with sceptics and bring them into the mainstream.

In the days immediately after the story broke, Curry had published her thoughts on the scandal at Steve McIntyre's *Climate Audit* blog. She appears to have been reasonably impressed with the explanations given by Gavin Schmidt, saying that these were 'helpful in providing explanations and the appropriate context', but it was also clear that she felt that, regardless of the explanations and alibis, there was still a considerable case for mainstream climatology to answer, and she took aim at climatologists' tendency to 'circle the wagons' rather than engaging their critics in rational debate. It appears that many mainstream climatologists, already angered by Curry's willingness to engage with sceptics, were highly unimpressed with her breaking of ranks and speaking out on Climategate. As she said in an interview with the *National Review*,

Somebody who was named in those e-mails e-mailed me and was rather upset about my lack of support and my speaking about this. [116]

She went on to note, however, that she had also received a considerable volume of emails in support of her position from scientists working outside the field of climatology.

DECEMBER – COPENHAGEN

The timing of the release of the Climategate emails is obscure, but it is possible that it was related to the then-imminent Copenhagen Climate Summit, at which world leaders were supposed to set out plans for mitigating global warming. This was no easy task, and in fact even months before the delegates arrived to start negotiations it had been clear that a meaningful agreement was highly unlikely. As the conference got under way, the revelations of the Climategate emails had gone viral on the

internet and it looked very much as if the damage done was already irreparable. Even so, there were signs that drastic steps were going to
be taken in a last-ditch attempt to bring the media frenzy under control. Shortly after the conference opened, Andy Revkin wrote a post on
his *New York Times* blog rounding up recent developments on the climate front, concentrating mostly on Climategate, but also mentioning a
mildly amusing story from Copenhagen. According to the German newspaper *Der Speigel*, the mayor of Copenhagen had advised visitors to the
conference to 'Be sustainable – don't buy sex' from the city's prostitutes.
When Revkin reported this, he received a threatening email from a climate scientist named Michael Schlesinger, who remonstrated with him
for mentioning the story and also for quoting two academics critical of
mainstream climate science – the father and son team Roger Pielke Sr
and Roger Pielke Jr.*

> Copenhagen prostitutes? Climate prostitutes? Shame on you for
> this gutter reportage. This is the second time this week I have
> written you thereon, the first about giving space in your blog to
> the Pielkes. The vibe that I am getting from here, there and ev
> erywhere is that your reportage is very worrisome to most climate
> scientists.
>
> Of course, your blog is your blog. But, I sense that you are about
> to experience the 'Big Cutoff' from those of us who believe we
> can no longer trust you, me included. Copenhagen prostitutes?
> Unbelievable and unacceptable. What are you doing and why? [117]

Revkin was also receiving pressure from *RealClimate* over his willingness to speak to the Pielkes, with a posting at around the same time
describing Revkin's decision to air Pielke Jnr's views as 'bizarre' and inviting him to make better use of his address book. [118]

If pressure on journalists to toe the line wasn't enough, the media
started to push a new line on the origin of the hacking. An article in the
Times questioned whether Climategate was the work of the Russian government, 'a state-sanctioned attempt to discredit the Copenhagen summit involving secret service espionage'. [119] The story became more eccentric still when Andrew Weaver, a prominent Canadian climatologist

---

*The Pielkes are not sceptics. Pielke Snr argues, however, that the IPCC understates
the impact of other manmade effects on the climate, such as land use. His son, a political
scientist, has worked in the areas of weather extremes, which he has shown to be unaffected by recent warming, and has also criticised policy responses to global warming.

from the University of Victoria, was reported as saying that there had been attempts to break into his office. The *Guardian* suggested that the breach 'heightened fears that climate-change deniers are mounting a campaign to discredit the work of leading meteorologists', an idea that was made to look rather ridiculous when, a few days later, it emerged that Weaver's university had notified staff that there had been attempts to steal IT equipment, not just from climatologists, but from a variety of university departments, including engineering and psychology.[120]

THE TRUTH BEHIND THE TRICK TO HIDE THE DECLINE

While mainstream climatology was concentrating on its public relations programme, sceptics were still trawling the emails for new information. In particular, McIntyre had turned his formidable analytical skills to discovering the precise details of the trick to hide the decline – what had happened and why it happened – and he published the full story in a series of postings at the time of the Copenhagen conference.

The trick to hide the decline had its roots in the period running up to the IPCC's Third Assessment Report in 2001, when the Hockey Stick had been catapulted to worldwide prominence. Over the years, this period had been thoroughly examined by McIntyre and his *Climate Audit* readers and many of the details were therefore familiar. As we have seen, the Hockey Stick suggested that, contrary to the popular conception, there had been no Medieval Warm Period. This was important because a lack of a medieval warming was a gift to the IPCC and environmentalists, who could issue dramatic press releases claiming that the modern warming was unprecedented in the last 1000 years.

The story begins back in the autumn of 1999, when the lead authors of the Third Assessment Report were to meet in Arusha, Tanzania, to discuss the Zero Order Draft of the report. Although there is no formal record of what took place at the meeting, in the aftermath the scientists involved exchanged emails in which they discussed what had happened and how to proceed, and many of these are preserved in the Climategate emails. On 22 September 1999, Chris Folland, a scientist at the UK's Met Office, wrote to Briffa, Mann and Jones about the paleoclimate chapter of the report. He told them that the intention was to have a diagram showing three temperature reconstructions on the same axes: Mann's Hockey Stick, Briffa's so-called MXD series and the reconstruction Jones had produced around the same time as Mann's paper (see Figure 4.1).

He also said that the graph was likely to be honoured by being included in the executive summary of the report – the so-called 'Summary for Policymakers'. But as Folland went on to explain, there was a significant problem: Briffa's paper was telling a very different story to the studies by Jones and Mann and was 'diluting the message'.

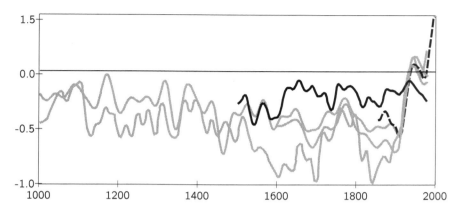

FIGURE 4.1: The Zero Order Draft graph

The instrumental temperatures are shown as a black dotted line at the right-hand side. Briffa's series is the other series shown in black, trending downwards at the right-hand side of the graph. The other paleoclimate series – the Hockey Stick, another Mann series and Jones' series – are in grey

> ...the [Briffa graph] with the tree-ring-only data somewhat contradicts [the Mann and Jones papers] and dilutes the message rather significantly. We want the truth. Mike thinks it lies nearer his result (which seems in accord with what we know about worldwide mountain glaciers and, less clearly, suspect about solar variations).[121]

What the scientists were discussing was the 'divergence problem', which, as we have seen, affected many tree ring series, including Briffa's Urals chronology, and threatened to undermine the whole of the field of paleoclimatology.* The Briffa series in the Zero Order Draft was the best-known example of the problem. As Figure 4.1 shows, the series has little or no long-term variation in its early sections, but also clearly exhibits a decline in the twentieth century, suggesting that these tree rings were not reliably tracking temperatures.

---

*See p. 56.

As Folland explained in his email to Mann, Jones and Briffa, dealing with the divergence problem was probably the most important issue for them to resolve on the paleoclimate chapter of the report. This appears to be because, behind the scenes, pressure was being applied to ensure that the authors came up with a particular result – a clear conclusion of unprecedented warmth. However, as Briffa noted, this was scarcely warranted, particularly in the light of the divergence problem.

> I know there is pressure to present a nice tidy story as regards 'apparent unprecedented warming in a thousand years or more in the proxy data' but in reality the situation is not quite so simple. We don't have a lot of proxies that come right up to date and [in] those that do (at least [in] a significant number of tree proxies) [there are] some unexpected changes in response that do not match the recent warming.[121]

As lead author on the relevant section of the report, Mann took the lead in emphasising the importance of dealing with the conflicting graphs, saying that the report would be stronger if they had a figure showing several mutually consistent temperature reconstructions...

> I believe strongly that the strength in our discussion will be the fact that certain key features of past climate estimates are robust among a number of quasi-independent and truly independent estimates, each of which is not without its own limitations and potential biases.[121]

...and he went on to discuss the concerns those at the Arusha meeting had had about the impact that inclusion of Briffa's series was likely to have:

> This is the problem we all picked up on (everyone in the room at IPCC was in agreement that this was a problem and a potential distraction/detraction from the reasonably consensus viewpoint we'd like to show [with respect to] the Jones *et al.* and Mann *et al.* series.[121]

As Mann explained, if they were to include Briffa's series then they would have to explain that 'something else' was causing the decline and he told the others that he was concerned that sceptics would 'have a field day' if scientists were to admit their ignorance. It would 'undermine faith in the paleoestimates'.[122] Various solutions to the problem seem to have been considered. Mann appears to have favoured leaving the Briffa study out altogether, while Jones appears to have floated the idea of having different diagrams in the summary and the chapter itself – presumably omitting Briffa's paper from the summary. McIntyre also came across evidence that shortly after the Arusha meeting someone – presumably Briffa – experimented with different ways of calibrating the MXD reconstruction, perhaps in order to make it match the Jones and Mann series better. [123]

Briffa appears to have been wavering, telling his colleagues that he did not think the current warming was unprecedented, but by the following day he seems to have been persuaded to toe the line and he sent a submissive email to Mann and Jones, explaining that he had been concerned over the talk of not muddying the waters – the IPCC was supposed to discuss differences of opinion and uncertainties. However, he said that he didn't want to overemphasise differences between the different reconstructions, and although we cannot see it in the emails, it appears that a way forward was agreed.

On 5 October 1999, Osborn sent Mann a new version of the Briffa series, which had been reprocessed to make it much more variable. This made it a much better match for the Jones and Mann reconstructions for most of its length, but had the unfortunate side effect of making the divergence much more prominent. This was clearly not going to be acceptable and so, for the final version of the graph, the divergence was simply deleted (see Figure 4.2). The decline had been hidden.

The truncation of the divergence problem from Briffa's series in the Third Assessment Report was *a* trick to hide the decline, but it was not the one referred to in the Climategate emails. The email in which Jones infamously referred to the trick to hide the decline appeared several weeks later and referred to a diagram Jones had prepared for the cover of a World Meteorological Organisation report. This was directly analogous to the diagram in the IPCC report, but with a slightly different approach to hiding the decline: instead of simply deleting the declining data, Jones replaced it with the instrumental figures and then smoothed the resulting series so that the join could no longer be seen. The decline was replaced by a rising trend, with the reader of the report none the wiser as to what had been done. The result can be seen in Figure 4.3 and the success in creating 'a nice tidy story' of 'unprecedented warming in a thousand years or more' is obvious.

By the time of the Fourth Assessment Report the difficulties with the divergence problem had still not been resolved, and Briffa had had the

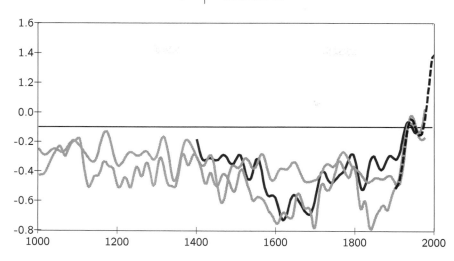

FIGURE 4.2: The decline hidden in the IPCC report.

The graphs represent temperature anomalies for each year. Briffa's series is the other series shown in black, truncated at 1960. The instrumental temperatures are the dotted line. The other paleoclimate series – the Hockey Stick and Jones' series – are in grey. The original figure appeared with uncertainty bands, which I have excluded from this emulation in order to simplify the presentation.

tricky problem of addressing it in the final report. It appears that his initial reaction was to try to repeat Mann's trick of simply deleting the decline without telling the report's readers what had been done. However, in the face of a blunt review comment from McIntyre he seems to have decided that he could not go through with the deception and added a lengthy paragraph to the report outlining the issue and its implications for the reliability of the reconstructions. However, he steadfastly refused to do anything about the truncation of the divergent data.

The Fourth Assessment Report version of the trick to hide the decline is shown in Figure 4.4(c), alongside the versions used in the Third Assessment Report and the WMO report. Figure 4.4(d) shows how the Fourth Assessment Report would have appeared had Briffa and Mann chosen not to hide the decline.

As should now be clear, hiding the decline was not, as Sarah Palin had said, an attempt to hide a cooling off in global temperatures; it was an attempt to hide the conflicting messages coming from the tree ring data, replacing a mixed message with a 'nice, tidy story' but one that was, however, quite false.

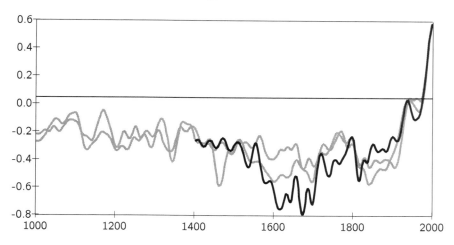

FIGURE 4.3: The trick to hide the decline: the WMO graph

Briffa's series is shown in black. With the instrumental temperatures spliced onto the end, the series now trends upwards at the right-hand side of the graph. The other paleoclimate series – the Hockey Stick and Jones' series – are in grey.

It is instructive to compare McIntyre's detailed analysis of the different tricks Mann and Jones had used to hide the decline to the explanations given by Schmidt at *RealClimate* and by Jones himself. As we have already seen, Jones had suggested to Ian Wishart that the instrumental data had been added on because paleoclimatologists 'don't always have the last few years' of data, a story that acknowledges the splicing of the two series, but which, in the light of McIntyre's analysis, appears to be grossly misleading, avoiding any mention of the deletion of the decline.[*] *RealClimate*, on the other hand, while acknowledging the existence of the divergence problem and the fact that it had been deleted, suggested that since Briffa had discussed the decline in his original paper, this somehow excused deleting it in the WMO and IPCC reports. They suggested that all Jones had done was to show the instrumental records 'alongside' the proxy data.[†] So while they acknowledged that Jones had deleted the divergence they completely ignored the splicing of the instrumental data, a very different version of events to that given by Jones.

---

[*]See p. 71.
[†]See p. 74.

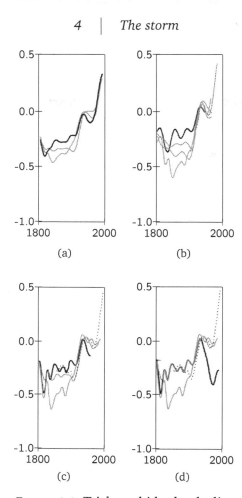

FIGURE 4.4: Tricks to hide the decline

Figures represented are temperature anomalies vs year. (a) The WMO graph; (b) the Third Assessment Report; (c) the Fourth Assessment Report; (d) the Fourth Assessment Report as it would have appeared without the truncation of Briffa's series. In each, Briffa's series is shown in black. The other paleoclimate series are in grey and the instrumental temperatures are the dotted lines.

Schmidt's suggestion that the decline was hidden 'in plain sight' was extraordinary since the decline was invisible to the readers of either report, and even if one of the policymakers reading it followed through the citation to Briffa's original paper, they would have been left none the wiser, since this paper would have shown them a completely different reconstruction to the one in the report – one with a decline rather than a rise in proxy values.

The new revelations about how the trick to hide the decline had been put together made an immediate impact, with the *Daily Mail* in London publishing an in-depth story on Climategate that examined the Trick in some detail.[124] It must have been very surprising for those inside CRU to see the intricacies of paleoclimate reconstructions discussed in the pages of a mid-market tabloid better known for its coverage of celebrity gossip than for science.

With new allegations still emerging from the emails and with media interest in the affair still showing no signs of waning, the politicians seem to have decided that they had to be seen to be acting. We have already met Phil Willis, the chairman of the House of Commons Science and Technology Committee, when he tried to explain away the emails as careless talk at the time the Climategate story first hit the headlines.* On 1 December 2009, Willis wrote to Professor Edward Acton, the UEA vice-chancellor, on behalf of the committee, asking for an explanation of what had happened and what the university intended to do about it. Somewhat more ominously, Willis also asked about allegations that CRU had deleted some of its station data. This was an issue that had shot to prominence a few days earlier, when it was reported in the *Sunday Times* that CRU no longer had the raw station data on which its important CRUTEM land temperature index was based.

> Scientists at the University of East Anglia (UEA) have admitted throwing away much of the raw temperature data on which their predictions of global warming are based.
>
> It means that other academics are not able to check basic calculations said to show a long-term rise in temperature over the past 150 years.
>
> The UEA's Climatic Research Unit (CRU) was forced to reveal the loss following requests for the data under Freedom of Information legislation.
>
> The data were gathered from weather stations around the world and then adjusted to take account of variables in the way they were collected. The revised figures were kept, but the originals – stored on paper and magnetic tape – were dumped to save space when the CRU moved to a new building.[125]

---

*See p. 79.

UEA's response to Willis' letter was swift and the same day the university issued a short press release announcing that Jones had decided to stand down until completion of an independent review into the allegations. The review, Jones said, would have his full support.[126]

Two days later, UEA issued another press release, announcing the appointment of a senior civil servant as head of the inquiry.[127] Sir Muir Russell was a former vice-chancellor of the University of Glasgow and now occupied the position of the Chairman of the Judicial Appointments Board for Scotland.*

The press release gave some brief details of the scope of the inquiry, and these were carefully examined by sceptics. It was noted, for example, that the inquiry would not be limited to just the hacked emails, but would also be able to look at 'other relevant e-mail exchanges' and any other information held at CRU. They would

> ...determine whether there is any evidence of the manipulation or suppression of data which is at odds with acceptable scientific practice and may therefore call into question any of the research outcomes.[127]

The panel would also be looking at the unit's compliance with 'best scientific practice' and FOI legislation, and its data management procedures. To onlookers this appeared all rather vague, but it at least seemed to be addressing the key areas of concern.

A few days later the university issued its formal response to Willis's letter.[128] This largely reiterated the points made in the press release – some details about how the emails had found their way onto the internet and information about the inquiry. But there were also some interesting details in response to Willis's other questions. In particular, Acton replied that 'None of the adjusted station data referred to in the emails that have been published has been destroyed'. To the parliamentarians, unversed in the intricacies of the CRU data, this may well have come over as a straightforward rebuttal. Those closer to the detail would of course have noticed that by making the 'adjusted data' the subject of his denial, Acton was neatly avoiding the actual accusation: that CRU had deleted some of the *raw* data on which the HADCRUT temperature index was based.

The Science and Technology Committee appear to have spent the Christmas break mulling over Acton's letter. On their return, and with

---

*He was, however, best known for his role in managing the construction of the Scottish Parliament building in Edinburgh, a project that had run disastrously over budget.

the controversy still raging in the media, they can have had little choice but to be seen to be trying to get to the bottom of the affair: a few weeks after returning from the winter recess, the committee announced their own inquiry into the Climategate emails. This was exciting news for the sceptic community, although the terms of reference for the inquiry were decidedly odd. The committee said that they wanted to look at the implications of the disclosures for the integrity of scientific research, the adequacy of the terms of reference for the UEA inquiry, and the question of whether CRU's global temperature series was independent of the other records around the globe.

None of these three areas of inquiry pointed clearly at the question of whether the allegations circulating on the internet were true or not, and the focus on the independence of the datasets in particular was very strange, as these were rather peripheral to the Climategate affair. In fact CRUTEM only received 25 mentions in the Climategate emails, most of which concerned paleoclimate reconstructions and the IPCC reports. There were no serious allegations about CRUTEM apart from those related to possible breaches of FOI legislation.* That said, the committee's focus on whether the terms of reference for the UEA inquiry were adequate was welcome, as it could be seen as a shot fired across the bows of the university, warning them not to whitewash the affair. This at least gave some encouragement to the sceptics. However, there was bad news to come.

In the wake of the announcement, a number of newspapers interviewed committee chairman Phil Willis to get some background on the committee's thinking and some of his responses suggested that he, at least, had strongly held views on climate change and about sceptics. In particular, Willis's remarks to the *Daily Telegraph* were highly controversial.

> There are a significant number of climate deniers, who are basically using the UEA emails to support the case this is poor science. We do not believe this is healthy and therefore we want to call in the UEA so that the public can see what they are saying.[129]

If Willis's earlier attempts to suggest that the contents of the Climategate emails were just loose talk had raised the suspicions of sceptics,

---

*It is now known that the suggestion that the committee focus on CRUTEM in this way came from one of its members, Evan Harris.

his use of the word 'denier' was highly alarming, being widely seen as a scurrilous attempt to compare the questioning of global warming science to holocaust denial. The situation looked bleaker still when the voting records of the committee members were examined and it was noted that the majority had been solidly behind most climate change legislation. However, the announcement of the inquiry indicated that submissions of evidence would be welcomed, so there would at least be a chance to set out the facts. The committee had also indicated that they would be taking oral evidence in March and, while it was not clear exactly who would be invited to speak, it seemed unlikely that UEA would not be called. The prospect of seeing CRU scientists cross-examined about the emails was a mouth-watering prospect for the sceptics, and one that could potentially undermine the credibility of mainstream climatology still further.

The news of the inquiry had only been public for a few hours before a potentially serious problem was pointed out. The UK's Labour government, in office since 1997, was then in its dying days and by law a general election had to have taken place by May 2010. The dissolution of Parliament ahead of the election campaign was expected imminently and there was considerable concern over whether there would be time for the Science and Technology Committee to consider the evidence, or perhaps even to hear evidence at all. But in the meantime there was nothing to do but to hope for the best.

THE SIDE INVESTIGATION

While the university had responded to the crisis by empanelling the Russell inquiry and had a parliamentary inquiry to deal with too, other prominent members of the scientific community had also developed an interest in events at UEA. Professor John Beddington was the government chief scientific advisor and turned out to have a central, if rather shadowy, role in some of the events that unfolded in the months that followed the Climategate disclosures. Some time in early December 2009, Beddington appears to have had a conversation with Professor Trevor Davies, the pro-vice chancellor for research at UEA and a former head of CRU. The nature of these exchanges is unclear and it is possible that there was a telephone conversation or a face-to-face meeting – there are apparently no written records. What is known about the exchanges is based on references to them in the minutes of some of the Russell panel's meetings.

The first of these meetings was in mid-December, when Russell met with Davies and a number of other senior UEA staff at the university. The minutes record that someone, apparently Davies, reported that Beddington had previously 'suggested that there could be a side investigation which looks at the science, in more detail'. [130] It was also noted that Russell would need to speak to Beddington, wording that suggests that the two men had not met previously. Thus it seems likely that it was Davies who had met Beddington a few weeks earlier. This surmise seems to be confirmed by the minutes of the meeting of Russell's team in the New Year, where he gave details of a meeting he had held with Beddington a few days earlier.

> This [meeting] had clarified what Sir John had in mind in terms of a 'side investigation'. It appeared that, following a conversation with Trevor Davies, pro-vc at UEA and a former head of CRU, Sir John saw advantage in an exercise to look at the CRU work without the 'adjustments' to data, of which much had been made by critics.

It seems clear then that Beddington, a man at the heart of government, was already deeply involved in the Climategate inquiries. Moreover, as the minutes make clear, he was doing much more than simply overseeing the process but was actively involved in determining the work plan of the panel – Beddington's intention appeared to be to shore up public perceptions of the the surface temperature records against the criticism of the sceptics. Whether Beddington was aware of just how peripheral the CRUTEM record was to the Climategate allegations is unclear, but the effect of his request was to shift the focus of the inquiry, at least in part, away from the actual allegations and onto areas where there was little expectation that significant issues would be found.*

A few days after that first mysterious conversation with Davies, at which the idea of a side inquiry had been raised, Davies wrote a strange letter to Beddington. The content was, of itself, somewhat uninteresting, simply setting down the university's understanding of what had happened and what steps were planned to rectify the situation. However, the letter was written without any reference to their earlier conversation – it is written almost as if this was the first contact between the two men. It is possible that this was just Davies setting things down in a formal manner, but equally it may be that Beddington wanted his involvement in defining the work of the inquiry to be kept as quiet as possible.

THE RUSSELL PANEL

By the start of February, Sir Muir Russell was ready for the public launch of the inquiry, and a press conference was arranged at the Science Media Centre in London to enable Russell to introduce his team. He would also be launching the inquiry's website, on which all submissions of evidence would be published and the findings revealed. Russell was keen to emphasise the inquiry's independence, although this claim rang slightly

---

*McIntyre, for example, had been saying for some time before Climategate that he did not expect a 'smoking gun' to eventually be found in the CRUTEM record - he thought the reason for CRU's obstruction of FOI requests was to protect their commercial interests and to hide a lack of quality control. [131]

hollow when he had to admit that its work was entirely funded by UEA itself. He also said that the university had had no influence on his choice of team members, which was interesting in the light of what happened next.

Just as when the Science and Technology Committee's hearings had been announced, sceptics had scrambled to assess the committee members' views on global warming, the announcement of the Russell panel was the signal for a rush to scour the web for information about the team chosen to investigate the Climategate emails (see Table 4.1). It was a matter of minutes before some startling facts emerged.

TABLE 4.1: Members of the Russell panel

Name	Background
Sir Muir Russell	Former vice-chancellor of Glasgow University
Prof. Geoffrey Boulton	Glaciologist, University of Edinburgh
Prof. Peter Clarke	Physicist, University of Edinburgh
David Eyton	Head of Research and Technology, British Petroleum
Prof. Jim Norton	IT expert
Dr Philip Campbell*	Editor of *Nature*; peer review expert

*Resigned on day of launch

One of the first panel members to attract attention was Philip Campbell, the editor-in-chief of the prestigious scientific journal, *Nature*. His appointment was immediately viewed with suspicion because it was *Nature* that had published the papers around which some of the most serious allegations were swirling. This meant that if the Russell panel found that the credibility of either of these papers was undermined in some way by the contents of the emails, it would probably reflect badly on the journal, or possibly even on Campbell himself, who had overseen the publication of the Hockey Stick paper and the rejection of McIntyre's explosive refutation of it, on the somewhat improbable grounds of lack of space.*

---

*See p. 13.

A second area of concern over Campbell's appointment was that *Nature* is widely seen by sceptics as a signed-up member of the environmental movement, regularly publishing aggressively worded denunciations of anyone who questioned the global warming orthodoxy. An editorial on Climategate, published a few weeks earlier, was headlined 'Stolen e-mails have revealed no scientific conspiracy' and declared that 'denialists use every means at their disposal to undermine trust in scientists and science'.[132] Although the editorial had no byline, the fact that such a strongly worded article appeared in his journal hardly suggested that Campbell was going to be an independent-minded panel member.

But the final straw came when a recording was unearthed of a brief interview Campbell had given to a Chinese radio station shortly after the Climategate affair had hit the media. In it Campbell appeared to make an unequivocal statement that the scientists involved were innocent:

> INTERVIEWER: I think you must have heard of the Climategate scandal recently. Some renowned global warming proponents showed a conspiracy to produce fraudulent data to support the global warming scenario. How do you see this scandal? Some say that this breaking couldn't come at a worse time because of the upcoming Copenhagen conference. What's your opinion?
>
> CAMPBELL: It's true that it comes at a bad time but it is not true that it is a scandal. The scientists have not hidden the data. If you look at the emails there [are] one or two bits of language that are jargon used between professionals that suggest something to outsiders that is wrong. In fact the only problem there has been is some official restriction on their ability to disseminate their data. Otherwise they have behaved as researchers should.
>
> INTERVIEWER: So you think there has been some misunderstanding between the scientists and the outsiders?
>
> CAMPBELL: Absolutely, absolutely.[133]

Given that one of the questions the panel was supposed to be investigating was whether CRU staff had 'hidden the data', Campbell had clearly prejudged the outcome of the investigation. The press conference finished at around midday on 11 February and by 2:30 pm the news of Campbell's *faux pas* was being put to the Russell team by the press. Shortly afterwards Campbell resigned from the inquiry, telling Channel Four News that he 'made the remarks in good faith on the basis of media reports of the leaks'. His departure was amusingly ascribed by the journalist concerned to a 'well-organised and highly-motivated campaign' by sceptics.[134]

Campbell's explanation is interesting given that he, or someone working closely with him, had similarly exonerated the CRU scientists in the editorial mentioned above. Although the editorial said that 'A fair reading of the e-mails reveals nothing to support the denialists' conspiracy theories' it seems likely that the journal had not in fact given a fair reading to the emails and that its position was at best based on little more than hearsay.

Before long a second member of Russell's team started to come in for some sustained scrutiny – the panel's only climatologist, Professor Geoffrey Boulton. Within minutes of his name being announced it was discovered that, earlier in his career, Boulton had been on the staff of the School of Environment at UEA for nearly twenty years – CRU, readers may remember, is a unit within this school. This meant that for several years Boulton had been a close colleague of Phil Jones and another central figure in the emails: Tom Wigley. Boulton now worked at the University of Edinburgh, where his office was almost next door to two other members of the Hockey Team: Tom Crowley and Gabriele Hegerl.

What was worse, it was also shown that Boulton was a regular and vocal commentator on the issue of global warming. In fact, just weeks before the Climategate disclosures, he had spoken at a meeting of senior scientists alongside Professor John Mitchell, the Met Office senior scientist, who was implicated in some of the allegations of wrongdoing arising from the emails. A precis of Boulton's speech gives a flavour of what he said:

> We have the evidence, we have a consensus on scientific interpretation, we have the investment, we know...that mitigation now rather than later is cheaper. But, we have not sorted out the politics and started to adapt behaviour to minimize risks. We cannot do this without public support. If we fail, we will be risking the consequences of catastrophic climate changes.[135]

Boulton's recent past was littered with further examples of his sounding the alarm on climate change – in 2008 he had even predicted that Himalayan glaciers would all have melted by 2050.[136] A remarkably similar claim was made by the IPCC in the Fourth Assessment Report – namely that the Himalayan glaciers would have melted by 2035. In fact though,

a more plausible figure is 2350,* and Boulton's mistake is therefore remarkable given that he is a glaciologist. Boulton also appeared to have
been involved in damage-limitation exercises in the wake of Climategate, having been a signatory of a public letter that had been organised
by the Met Office in the wake of Climategate, which sought to defend the
integrity of climate science. He had also authorised the writing of a briefing paper for the Royal Society of Edinburgh that had included a hockey
stick diagram, and which claimed incorrectly that this reconstruction was
'independent of' the hide the decline graph. [137,138]

What made Boulton's presence on the panel even worse was that the
new Russell inquiry website included clear statements on panel members'
views on climate change that made it quite clear that he was unsuitable.
In the 'About the Review' section, it was said that 'none [of the panel
members] have any links to the Climatic Research Unit, or the... IPCC', a
position that was repeated in the FAQ section:

> *Do any of the review team members have a predetermined view on
> climate change and climate science?*
>
> No. Members of the research team come from a variety of scien-
> tific backgrounds. They were selected on the basis they have no
> prejudicial interest in climate change and climate science and for
> the contribution they can make to the issues the Review is looking
> at.[†]

Boulton had even gone out of his way to emphasise his distance from
the climatology community in the press conference, where he is reported
as saying 'I am not involved in... issues of recent and current climate
nor am I part of that community'.[140] If the contradiction between the
panel's stated position and the facts were not bad enough, Boulton then
admitted in an interview with the *Times* that he *was* part of the climatology community and *did* have strong views on global warming, saying
that this was an inevitable consequence of the need for the committee to
have a member with a background in climatology:

> I may be rapped over the knuckles by Sir Muir for saying this, but
> I think that statement needs to be clarified. I think the committee

---

*The IPCC claim was at the centre of a furore over uncorrected mistakes in the Fourth
Assessment Report at around this time.

†The FAQ section has now been removed from the website, but the words can be seen
in several of the submissions of evidence to the Russell inquiry.[139]

needs someone like me who is close to the field of climate change and it would be quite amazing if that person didn't have a view on one side or the other.[141]

Boulton's presence on the panel appeared indefensible. However, Russell appears to have decided that having lost Campbell, he could not afford to lose another member of his team, and despite a wave of criticism he brazened it out and Boulton kept his place on the team. Russell's only fig leaf of a defence was to claim that it was impossible to find somebody with the required qualifications and experience who had not formed an opinion on climate change,[142] an argument that appeared to completely contradict his earlier position that none of the panellists had such firm opinions.

Russell's willingness to sacrifice the panel's credibility in order to keep Boulton immediately raised question marks over why the glaciologist should survive when Campbell, whose position was arguably rather less compromised, should have been forced out. Of course, it was quite possible that a second resignation would have caused the collapse of the entire inquiry, but it was also noted that the panel was to hold its meetings at the Royal Society of Edinburgh, where Boulton was general secretary – in essence the chief executive. The inquiry secretariat would also be staffed with RSE personnel. These observations in themselves were circumstantial, but McIntyre's subsequent discovery that the issues paper had been prepared by Boulton rather than Russell, and the later revelation that Boulton had been in charge of contacting potential team members* did start to make it seem as if it was Boulton who was running the show rather than Russell.[143]

With Boulton refusing to stand down, the obvious step was to raise the issue with the Science and Technology Committee, whose remit, as we have seen, covered the adequacy of the UEA inquiry. However, in practice this was no longer possible because by an extraordinary coincidence, Russell had chosen to announce his team just a day after the closing date for submissions to Parliament. Suddenly it was possible to see why there had been such a long delay between the announcement of the Russell inquiry on 3 December 2009 and the announcement of the panel members on 11 February 2010. It was hard to let go of the suspicion that

---

*Boulton's recruitment of the other panel members is surmised from his expense claims, which record that in late December he was 'advising Sir M-R on remit, science, and personnel for review team, contacting potential members...

the Russell team had planned the timing of their press conference very carefully.

With the forensic analysis of the panel members complete, attention turned to the Russell panel's website and, in particular, to the programme of work and the issues paper that accompanied it.[144,145] The issues paper in particular was a source of concern and McIntyre was singularly unimpressed. For a start the panel admitted that they had prepared the issues paper before even completing their reading of the emails. Given this failure, their previous undertakings* to read email correspondence beyond the Climategate dossier now looked as though they would turn out to be worthless. The panel's failure to get to grips with their brief was made obvious at their press conference, when Boulton had appeared to confuse the notorious 'delete all emails' request with a completely different message in which Jones had said that he would keep one of McKitrick's papers out of the IPCC reports. Few people can have formed a favourable impression of the panel.

THE SCIENTIFIC APPRAISAL PANEL

On the same day as Russell's press conference, UEA made a further announcement: they were going to split responsibility for the investigation of CRU. Russell's team would now only be looking at the conduct of the scientists while a new panel would look at the science:

> *New scientific assessment of climatic research publications announced*
>
> An independent external reappraisal of the science in the Climatic Research Unit's... key publications has been announced by the University of East Anglia.
>
> The Royal Society will assist the University in identifying assessors with the requisite expertise, standing and independence.[146]

The appearance of the new panel was slightly odd, as many of the allegations that Russell had said he was going to investigate were highly technical, but the Russell panel's terms of reference made it clear that assessing the science was not part of their job – that was for the new inquiry:

> The review's remit does not invite it to re-appraise the scientific work of CRU. That re-appraisal is being separately commissioned by UEA, with the assistance of the Royal Society.

*See p. 91.

These words originally appeared on the Russell panel's website and were quoted at the time in the comments at *Climate Audit*.[147] Subsequently, the text appears to have been changed to:

> The Review's remit does *not* invite it to re-do the scientific work of CRU. An audit or assessment of that type would be a different exercise, requiring different skills and resources. However, the review's conclusions will be useful to any such audit by pointing to any steps that need to be taken in relation to data, its availability and its handling.[144]

Another concern arising from the Russell panel's announcement was the timescales for members of the public to make submissions. While it had taken UEA the best part of three months to put together the inquiry team, evidence to the inquiry had to be submitted by the end of the month – there were only eleven working days for over 1000 emails and code to be summarised into a form suitable for submission, a task which was almost impossible.* The panel was clearly expecting to be criticised for imposing such a tight timetable and had outlined an unconvincing explanation as to why it was necessary:

> Everyone likely to make a submission to the review already has a view on the issues within its remit, and the hacked emails are in the public domain. The review team is also keen to ensure that preliminary conclusions are presented to UEA in the spring, as has been requested.†

Clearly, having a view on the issues was unlikely to make much difference to people's ability to summarise them all in writing in such a short space of time, but there was nothing to be done about it. Those who wished to express their concerns about the conduct of the CRU scientists would have to cut their submissions short as best they could.

---

*The Russell panel's report says that submissions of evidence were accepted up to 16 April, some six weeks after the deadline that had originally been announced.

†The page of the Russell panel website that contained this information has since been removed, but the relevant part was quoted at *Climate Audit*. See the comment by 'hro001' at 11:54pm on 11 February 2010.[147]

CHAPTER 5

SAFEKEEPING

THE INQUIRY BEGINS

Although we have already seen the disastrous circumstances surrounding the launch of his inquiry, in fact Russell had been hard at work behind the scenes since his appointment in December. It is important to examine his activity during this period since it explains much of what followed.

Within days of the announcement of his appointment as head of the inquiry into the emails, Russell made his first visit to UEA. His meetings, which took place on 18 December 2009, appear to have been largely concerned with preparations for the inquiry and thus to be mostly insignificant. However, although the records are thin, there are some intriguing details of what was discussed.

Russell's first engagement was with Trevor Davies, the university's pro-vice chancellor for research,* and other senior members of staff. The minutes of their meeting show that Davies was still very much taking the lead for the university, confirming the impression given by his earlier correspondence with Beddington.

One question that was addressed by Russell and Davies was the recruitment of panel members, but at that stage no decisions had been made. However, the minutes reveal some of the candidates who were being considered, and some of these were quite surprising. Professor Bob Watson, for example, was the chief scientist at DEFRA and had been the head of the IPCC when it had introduced the Hockey Stick graph as a sales tool for the global warming movement. Watson was also a former employee of UEA, where he still held a part-time position. It is therefore extraordinary that his name should have been considered, but fortunately good sense appears to have prevailed and nothing more was heard of the idea. [130]

Several of the meetings that day touched on the subject of how the emails had reached the public realm. Russell had asked the question directly but had received no clear answer – according to the university registrar, the police were involved, as were some external consultants, but the investigation was expected to be long and complex. The IT staff gave

---

*We met Davies in Chapter 4 – see p. 93.

some more details: in the first days after the emails had reached the internet, CRU's IT administrator had performed a preliminary investigation and had concluded that there had been a sophisticated hacking, but by the time of Russell's visit it seems that firm conclusions had proved elusive, with one of the attendees at the meeting referring to a 'hack/leak'. Another mentioned that the police were looking at both possibilities and it was noted that an 'inside job' was much more likely.[148] These positions were surprising given that in its public pronouncements the university had been unequivocal that they had been the victim of a hack.[127]

Jonathan Colam-French, the university's IT director, told Russell that he believed the BBC had been offered sight of the emails in October, and said that this suggested that the hacker had accessed CRU's data more than once. His source for this story was almost certainly a rumour that was circulating on the internet at the time. Shortly before Climategate, Paul Hudson, a somewhat sceptical BBC weather correspondent, had written an article about the recent lack of warming in global temperature records, suggesting that some cooling might be expected in future decades. There had been consternation over Hudson's article among members of the Hockey Team, including Mann, Schneider, Trenberth and Wigley, and in due course the correspondence between them was forwarded to Hudson in order to put pressure on him to amend his position.

After the disclosure of the Climategate emails, the messages that mentioned Hudson had become the subject of considerable discussion on the internet and he had been forced to write a short article on his blog to explain what had happened. However, he worded his story somewhat vaguely, telling his readers that he had been 'forwarded the chain of emails'.[149] Unfortunately, his words were misinterpreted, and were quickly taken by many to be an admission that he had been offered the entire Climategate database but had spurned the offer of a 'scoop' in order to protect the BBC's position as promoters of the global warming orthodoxy. Despite a subsequent clarification, the story by then had legs and Hudson's innocent involvement has never been completely accepted in some quarters.[150] It is nevertheless surprising that, nearly a month after Hudson had set the record straight, UEA still seemed to be labouring under a misapprehension about what had happened.

Russell's meetings with the scientists, on the other hand, seem to have been relatively inconsequential.[151,152] He had been told by Davies that Jones and Briffa were under a great deal of pressure, and when he met the subjects of his investigation, Davies was on hand as a chaperone. Much of this first meeting appears to have been taken up with some superficial discussion of CRU's science. There was some discussion of the uncertainties involved in CRU's temperature reconstructions, and Jones told the meeting that even the scientists didn't understand the error bars in the kinds of research done at CRU. He also asserted that bloggers didn't understand the science or peer review.

With peer review being at the centre of Russell's investigation, it is not surprising that Russell picked Jones up on this last point, although in a rather surprising way. Russell's remit required him to investigate whether the peer review system had been corrupted or undermined, yet according to the minutes, Russell raised the possibility of whether it would be possible to recruit a panel member who would 'support/explain the robustness of the peer review process'.[151] He went on to suggest that the editor of *Nature* – Philip Campbell – might be a suitable candidate because of his tendency to 'publish rebuttals and criticisms'.* There was, according to Russell, 'an attack on peer review which it would be helpful to address'. Remarkably then, he appears to have seen his task as defending the reputation of peer review rather than finding out if Jones and his colleagues had undermined it. The minutes record that everyone thought that Russell's ideas were good ones.

THE INFORMATION COMMISSIONER

While Russell was busy at East Anglia, sceptics were also hard at work. On that first Monday morning when the Climategate story broke in the media, David Holland had quickly realised the importance of what he was reading. Back in 2008 he had issued a formal appeal to the UK's Information Commissioner's Office (ICO) over the university's refusal to release Briffa's IPCC correspondence. The commissioner is the official charged with enforcing the provisions of FOI legislation and has the power to compel public bodies to release information.

As Holland had explained in his complaint, UEA's claim that the IPCC review was subject to confidentiality considerations flew in the face of the IPCC's own guiding rules: 'comprehensiveness, objectivity, openness and transparency'. Now, however, with clear evidence that the scientists within CRU had conspired to block his FOI requests, his case suddenly

---

*We have seen one example of the kind of editorial published by *Nature* earlier in the story – see p. 97.

looked much stronger. Moreover, Holland was well aware how serious things were getting. Deleting emails subject to an FOI request is a criminal offence under UK law, so if any of the CRU scientists were found to have done so – something that appeared likely from several of the emails – it could feasibly have spelt the end of their careers.

Whether Jones or Briffa had in fact deleted anything was unclear at that time – the emails in which Jones discussed using memory sticks were not yet public.* However, as we have seen, Jones had privately told colleagues that he had 'deleted loads of emails', noting shortly afterwards that he had been told by the university FOI officer that he shouldn't be doing this unless it was in the normal course of maintaining his email box.†153 In the wake of the Climategate disclosures, Jones had slightly altered this story, giving a journalist an explanation that appears to have been close to the truth:

> We haven't deleted any emails. I delete my own personal emails a year at a time regardless of subject as I have too many, but the university still has the emails.154

Their potential breaches of FOI legislation were not the only problem facing the CRU scientists. Frustrated by his inability to extract Briffa's IPCC correspondence from CRU, Holland had issued a further request, this time for all the information that UEA held about him, in the hope that this might reveal compromising information about the activities of the CRU scientists. At the time, the university had said that they held nothing beyond Holland's various FOI requests, a claim that was now shown to be false by the Climategate emails, which contained several messages about him. Since UEA would have had to consider these new requests under the UK's Data Protection Act, the withholding of this correspondence meant that the university was potentially in breach of this second piece of legislation too.

With UEA staff now likely to be guilty of multiple breaches of FOI legislation, Holland wasted no time in putting together a letter to the Commissioner, describing the contents of the emails and outlining how they threw further light on his original complaint. With stories of deletion of data and emails all over the media, the Commissioner had no choice

---

*A second batch of emails was released in November 2011.
†See p. 53.

but to respond and a short time later contacted Holland to arrange a meeting. The fourth inquiry into Climategate was under way.*

The meeting took place in mid-December at the Information Commission offices in Cheshire. As Holland describes it, the meeting was somewhat surreal, with the commissioner asking him to put together a detailed statement setting out all the detail of the emails as they related to his complaint. This seemed strange, since the emails were readily available on the internet and, in his letter, Holland had already set out plenty of areas for the commissioner to look at. However, reluctantly, he agreed to do as he was asked and a second meeting was arranged for the New Year.

Meanwhile, the commissioner had some bad news to relay. Although the law allowed for prosecutions for deliberate breaches of the FOI legislation, any such actions had to take place in a magistrate's court, and the Magistrates Act decreed that a statute of limitations of only six months should apply. Since any deletion of emails was likely to have taken place over a year earlier, it was apparent that nobody from CRU was going to answer their case in a court of law.

Interestingly, the near-impossibility of bringing a prosecution under the FOI laws had been recognised even before Climategate. Freedom of information campaigners had realised that investigations of complaints were likely to be long, drawn-out affairs.† Because of this, even wholesale shredding of documents by civil servants would go unpunished because of the statute of limitations in the magistrates' courts. In fact, this statute of limitations has been a concern in the framing of several pieces of UK legislation, with legislators writing in specific extensions as required. An amendment had even been tabled in Parliament to bring FOI laws into line, but remarkably the government had rejected the closing of this obvious loophole, claiming that there was 'no evidence... that the current six-month time limit presents a systemic problem for the Information Commissioner or any other prosecutor'.[155] This very much appeared to open the way for breaches of the law to go unpunished.

The commissioner's news was very disappointing for Holland, but nevertheless finding the truth was a higher priority than getting a prosecution, and he returned home to set out his case in more detail.

---

*The Information Commissioner's inquiry was actually the first to get under way. The narrative structure of the Climategate story has made it more convenient to mention it later than the other inquiries.

†It normally takes the ICO more than six months to even begin an investigation.

While Holland was trying to persuade the ICO to take his complaint seriously, Russell was holding discussions with Beddington in order to clarify the issue of the 'side investigation', with the chief scientist suggesting that Russell's team should try to rework the CRUTEM record with and without CRU's adjustments. Beddington apparently believed that there would be little difference between the raw and corrected data, and he and Davies felt that this would 'would be helpful in responding to criticisms'.[156] The focus on responding to criticisms is a worrying one, strongly suggesting that Beddington saw his role as one of defending CRU rather than ensuring that a proper investigation was carried out.

Beddington's focus on the temperature records is also strikingly reminiscent of the Science and Technology Committee who, as we have seen, had also decided to devote a considerable proportion of their inquiry to this peripheral aspect of the Climategate emails.* It is possible that there had been communication, either direct or otherwise, between Beddington and members of the committee. In the event the Russell panel rejected the raw/adjusted data comparison proposed by Beddington, although the reasons why are somewhat unclear – the minutes suggest that the panel saw more merit in looking at sensitivity analyses. But either way, Beddington's intervention looks important: its effect was to divert the panel's attention away from the serious issues that needed to be examined. And as we will see, the Russell panel's investigations of some of the most important Climategate allegations were either dramatically curtailed or were not performed at all.

MEETING THE COMMISSIONER

At the end of January, Russell was on site at UEA again, this time accompanied by Jim Norton, the inquiry's newly appointed computer expert.†
The two men were to hold meetings with Norfolk Constabulary and representatives of the Information Commissioner and in the event all three groups attended both meetings. In addition they would also meet with Jones and Briffa to discuss the temperature records.

---

*The announcement of the parliamentary hearings was actually to take place ten days later, but I have reordered the sequence to simplify the narrative.

†Norton is strictly *Professor* Jim Norton, holding a visiting professorship at the University of Sheffield. He is not, however, an academic, describing himself as 'an independent director and policy advisor'.

With the possibility of a criminal breach in the law to be considered, FOI was already one of the most important areas of Russell's investigation, but the need to investigate had been raised still further by some extraordinary information that Russell had received from UEA's IT director, Jonathan Colam-French, at their meeting the previous month. Colam-French had been discussing the way data was managed at CRU and had explained that procedures were somewhat chaotic, with data and emails often stored on the hard discs of staff members rather than on a central server. As an example of this kind of thing, Colam-French volunteered the information that Keith Briffa had taken home 'emails that were subject to FOI to ensure their safekeeping'.[148] The emails in question could only be those that had been requested by Holland, and so the truth was now out in the open – Briffa had taken a copy of his emails on a removable medium and then deleted the originals from his hard drive. He could presumably then have refused any FOI request on the grounds that the information was not held.

The messages in which Jones admitted using memory sticks in the same way did not come to light until two years later, when a second batch of UEA emails was released onto the internet, so for the time being it was only Briffa's transgression that had been revealed to the inquiry. But when the legislation was framed, this kind of behaviour had been foreseen, and it is a criminal offence to 'block' FOI requests in this way. Russell now knew that Briffa had breached FOI legislation.

At this point it is important to assess who knew about Briffa's 'safekeeping' of the IPCC emails and to consider how this news was disseminated to other members of the Russell inquiry and to senior staff at UEA. The only attendees at the meeting in December had been the university IT staff, Lisa Williams, an assistant registrar from Acton's office who was responsible for taking the minutes, and Russell himself. It is therefore hard to credit the idea that none of these people would have relayed the news to Acton and Davies. However, now that he was at the university again, Russell was at least going to be able to bring the ICO into the picture.

The minutes of the January meeting between Russell and Norton with the ICO – with officers from the Norfolk police in attendance – are therefore of considerable interest.[157] The details of the ICO's investigation were apparently discussed, and it was noted that 'from their analysis of the information it was reasonable to conclude that an offence...may have been committed'. In fact the commissioner was confident that breaches of both the FOI Act and the Data Protection Act could be demonstrated. UEA's defence of their conduct appears to be that the responses they had given Holland were 'correct' at the time, because the university FOI staff were unaware of the existence of the backup server that had been identified as the source of the Climategate disclosures. The excuse appeared flimsy, but despite this, it was also made clear that nobody at UEA would be held accountable – the statute of limitations was to stand in the way of any prosecution.*

However, despite containing many interesting facts about the investigations of Russell and the ICO, the most important feature of the minutes of the meeting are what they do not contain: there is no mention of Briffa or of 'safekeeping'.† This is a remarkable omission.

QUESTIONING THE SCIENTISTS

As part of the same visit, Russell and Norton met with Jones and Briffa for what was described in the minutes as an 'exploratory' interview. The meeting began with Russell briefing the two scientists on the nature of the inquiry, telling them that

> [it] would focus on the scientific rigour with which data had been collected, processed and presented in the context of the scientific norms relevant at the time... [158]

...a formulation that seemed to suggest all sorts of escape routes. Russell also said the review would look at the 'extent and effectiveness' of the peer review process. This also appeared an odd form of words, seeming to tiptoe around the actual allegation, namely that Jones had been involved in *undermining* the peer review process. It is certainly clear from the minutes that neither Russell nor Norton questioned Jones or Briffa on the specific allegations made in this area. The complete record of the conversation on this subject is as follows:

Peer review

---

*It is worth noting, however, that concealment of emails through use of memory sticks might well have represented an ongoing offence, to which the statute of limitations would not apply. If this is the case then the ICO's failure to prosecute is inexplicable.

†David Holland issued an FOI request for the ICO's records of the meeting and was told that they held no relevant information. An officer on the ICO's investigation team said he could not recall having been told about the use of memory sticks. It is perhaps also worthy of comment that the Russell panel's minutes are dated 25 February 2010, more than a month after the actual meeting took place.

Both professors were questioned as to how effective peer review could be carried out on papers when there appeared to be no effective standards for the development and manipulation of the base data sets and no independent verification of processing integrity. There was a general recognition that work needed to be done to create a more effective structure for peer review.

Secretary's note: Subsequently, Profs. Briffa and Jones outlined that effective peer review of the papers published by CRU is carried out by the journals that publish the papers CRU personnel write. This is in common with all scientists working in climate science and all other fields. There are also internal informal review mechanisms within CRU, where staff read and critique each others' papers prior to submission.

This was the only time that the interview touched upon one of the main allegations, and then only in a way that avoided discussion of the actual accusation. However, once again the most remarkable aspect of the interview is that Russell appears to have made no attempt to question Briffa about Colam-French's allegation that emails subject to FOI requests had been taken home 'for safekeeping'. This omission seems even more culpable in light of the fact that this was the last time that Russell was to meet Jones or Briffa. Despite being the head of the inquiry, he decided not to involve himself in any of the subsequent interviews of the two scientists, handing over responsibility for this task to other members of the panel. The implication of this remarkable decision is that, subsequent to the launch of the review, then still two weeks away, Russell did not meet any of the principal actors in the Climategate affair. And since Russell was the only member of the review team who we know was aware of Colam-French's revelations about Briffa's 'safekeeping' of the emails,* it is possible that even before the launch, a decision had been taken to quietly overlook what Colam-French had revealed.

ACTON'S TRICK

The news of the time-barring of prosecutions soon seeped out into the media, and journalists scenting a story began to check out exactly what was known and what was not. In particular, the *Sunday Times'* science correspondent Jonathan Leake made contact with the Information Commissioner's office to find out why nobody was going to be brought to

---

*There is no indication in the minutes that Russell and his team discussed the issue subsequently.

justice. Leake appears to have been persistent, and eventually obtained a statement that clarified the position regarding the statute of limitations.[159] The ICO explained that they were essentially pursuing two separate inquiries. One was to investigate Holland's original appeal against UEA's refusal to release the IPCC emails, and this investigation was ongoing. The other was the investigation of the deliberate blocking of requests by deleting data:

> The emails which are now public reveal that Mr Holland's requests under the Freedom of Information Act were not dealt with as they should have been under the legislation. Section 77 of the Freedom of Information Act makes it an offence for public authorities to act so as to prevent intentionally the disclosure of requested information. Mr Holland's FOI requests were submitted in 2007/8, but it has only recently come to light that they were not dealt with in accordance with the Act.

> The legislation requires action within six months of the offence taking place, so by the time the action taken came to light the opportunity to consider a prosecution was long gone. The ICO is gathering evidence from this and other time-barred cases to support the case for a change in the law. It is important to note that the ICO enforces the law as it stands – we do not make it.[159]

So it was clear that in the Commissioner's mind it was highly likely that a breach of the law had taken place. However, there was dismay among the sceptic community that nobody would be brought to justice as a result, and there was considerable discussion of how the law could have been so badly framed and whether an alternative prosecution would be possible – one theory that gained some traction was that conspiracy laws might also have been breached, but this idea was soon quashed. But as Holland had learned a month earlier, it appeared clear that the CRU scientists were going to get away with it. It was to be the first of many disappointments.

Throughout February, submissions of evidence started to flow to the Science and Technology Committee, setting out the case for and against the CRU scientists. Among these was a contribution from UEA itself, part of which was an attack on the commissioner's statement about his concerns over the university's compliance with FOI legislation.

> On 22 January 2010, the Information Commissioner's Office (ICO) released a statement to a journalist, which was widely misinterpreted in the media as a finding by the ICO that UEA had breached

Section 77 of the FOIA by withholding raw data. A subsequent letter to UEA from the ICO (29 January 2010) indicated that no breach of the law has been established; that the evidence the ICO had in mind about whether there was a breach was no more than prima facie; and that the FOI request at issue did not concern raw data but private email exchanges.[160]

The UEA submission was delivered to the committee on 25 February, almost a month after the ICO's statement to Leake and, more significantly, just a week after Russell had been told about Briffa's safekeeping of the IPCC emails. The university's decision to respond in these terms is therefore very surprising. As noted above, it must be considered extremely unlikely that Colam-French had not told UEA's vice-chancellor Edward Acton about what Briffa had done. Acton's decision to defend Briffa in the terms he did appears inexplicable.

The following day, the university issued a somewhat opaque public statement, which expanded on the story of their dealings with the ICO.[161] Although what prompted this second step is not clear, it appears that after UEA's evidence to the Science and Technology Committee had become public, the ICO had taken exception to the university's story. UEA were standing their ground and the two sides were apparently at loggerheads, but they seem to have agreed to publish their correspondence so that everyone could assess their positions. UEA's case was that the ICO had not investigated the allegations of FOI offences any further and that a breach had not therefore been established. The university continued:

> The University regrets that the ICO statement has been interpreted by some to indicate that such an investigation has already been held and is pleased to set the record straight. The existence or otherwise of prima facie evidence is insufficient to reach any conclusions about this matter.[161]

The Commissioner, however, had strongly resisted an attempt by Acton to get him to water the statement down and made it abundantly clear that the evidence already to hand was extremely strong:

> The prima facie evidence from the published emails indicate an attempt to defeat disclosure by deleting information. It is hard to imagine more cogent prima facie evidence... In the event the matter cannot be taken forward because of the statutory time limit...

> The fact that the elements of a section 77 offence may have been
> found here, but cannot be acted on because of the elapsed time, is
> a very serious matter. [162]

It is perhaps significant that the ICO refers to the evidence 'from the
published emails'. This would appear to suggest that they were unaware
of Briffa's safekeeping of the IPCC correspondence, strengthening the impression already given by the minutes of Russell's meeting with the ICO
that the commissioner had not been told about this damning piece of evidence. It is hard to imagine why Russell would have chosen not to tell
the commissioner this important news, although the fact that the commissioner had already announced that no prosecutions would be possible
may be significant.

From their separate versions of events, it appears that the commissioner and the university agreed on the facts. However, Acton needed
to spin furiously to put the best possible gloss on them. He therefore
emphasised the prima facie nature of the evidence, neatly changing the
Commissioner's 'cogent prima facie evidence' into evidence that was 'no
more than prima facie'. At the same time, he inserted the word 'private' before 'emails' – a view of the Climategate correspondence that had
not appeared in the commissioner's statement to Leake. Acton was once
again trying to suggest that the CRU scientists were the aggrieved parties
in the Climategate affair; Steve McIntyre called it 'a trick':

> . . . Acton resorted to a 'trick' – what Gavin Schmidt and the climate
> science 'community' call a 'good way' of dealing with a problem.
> Acton's trick has blown up on him with the release of the actual
> ICO letter. I think that the 'community' would have concluded by
> now that the public is fed up with 'tricks' from the 'community'. [163]

Unfortunately for Acton, his actions had not gone unnoticed on the
Science and Technology Committee either. In a newspaper report that
appeared shortly after the correspondence between Acton and the Commissioner was made public, one of the committee's most prominent members, Dr Evan Harris, was quoted as saying:

> It seems unwise, at best, for the University of East Anglia to at-
> tempt to portray a letter from the Information Commissioner's Of-
> fice in a good light, in evidence to the select committee, because it
> is inevitable that the committee will find that letter, and notice any
> discrepancy.

It would be a wiser course for the university not to provide any suspicion that they might be seeking to enable the wrong impression to be gained. [162]

CHAPTER 6

THE PENN STATE INQUIRY

Within days of the release of the Climategate emails, Michael Mann's employer, Penn State University, had announced that it was going to launch an inquiry into the allegations that were being made on the internet about his conduct. The press release was short and to the point, making it abundantly clear that this was to be a private affair.

> In recent days a lengthy file of emails has been made public... The University is looking into this matter... following a well defined policy used in such cases. No public discussion of the matter will occur while the University is reviewing the concerns that have been raised. [164]

Sceptics were understandably excited by the news, but Mann in fact appeared to be quite unruffled by the attention, and responded to journalists' questions with his customary bravado, declaring that he was 'very happy' that the university was going to look into the allegations. He was quite clear that while some of the emails might have looked bad, his behaviour had been blameless, as the Penn State student newspaper reported:

> [Although] he says he was asked to delete selected e-mails by Jones, Mann said he did not comply with the request. He does not believe any of his colleagues went through with the deletion either. [165]

Despite the university's wish for secrecy, over the following weeks some limited information began to emerge about the inquiry. In common with most university inquiries of this kind there was to be a two-step process. The first part was the 'inquiry' – something equivalent to a grand jury hearing – where a three-member panel would decide if there was a case for Mann to answer. If the inquiry decided it was necessary, a larger team – the investigation – would determine Mann's guilt or innocence. [166]

When the names of the inquiry committee panel members were announced, there were no great surprises. There was Henry Foley, the Vice President for Research and Dean of the Graduate School, Alan Scaroni, the Associate Dean for Graduate Education and Research at the College of Earth and Mineral Sciences, and Candice Yekel, the university's Research Integrity Officer.* William Brune, Mann's head of department, would also be on hand as a consultant to the inquiry, although he would not take part in the decision-making process. The involvement of Scaroni was significant, since he was standing in for William Easterling, the Dean, who had 'recused' himself 'for personal reasons'.[167]

THE INQUIRY

By the beginning of February, word had emerged that the inquiry had nearly completed its work and a few days later the findings were published.[167] The results were, as expected, an almost complete victory for Mann. However, concerns were soon being raised as to just how diligent the inquiry had been.

For reasons that were not entirely clear, the inquiry team had not asked for submissions of evidence and had therefore chosen to try to determine the scope of their work by reference to the allegations circulating on the internet and to Mann's email correspondence. Surprisingly, however, they had chosen to limit their reading to those of his emails that appeared in the Climategate disclosures, failing to examine any of the rest of his correspondence.

The allegations that the panel had eventually formulated were listed under four headings: suppression and falsification of data, deletion of emails and data, misuse of confidential information and deviation from accepted academic standards. These could be construed as covering the main areas of concern, although the wording was extremely vague in places. For example, would the allegation that Mann had threatened journals be adequately covered by an investigation into compliance with 'accepted academic standards'? It was hard to tell, but the rest of the report would presumably give a clear answer.

According to the report, the inquiry team had interviewed Mann for two hours and they noted that he had answered all of the allegations. In particular he said that he had not falsified or manipulated data, he had not deleted emails, and had not engaged in activities that were inconsistent with accepted standards in the scholarly community. This was very much as expected, but some of these details as reported by the in-

---

*The inquiry was initially under the direction of Eva Pell, the Senior Vice-President for Research and the Dean of Graduate Studies, but she left the university soon afterwards. Foley worked ex-officio on the inquiry until she departed.

quiry team suggest that the questions and answers were rather missing the point. For example, Mann had claimed that he had 'never used inappropriate influence in reviewing papers by other scientists who disagreed with the conclusions of his science', a position that rather avoided the precise allegation – that he might have threatened journal editors who published contrary opinions. The words 'in reviewing' papers also made a nonsense of his denial, since in neither the Soon and Baliunas affair nor the Saiers affair was Mann one of the reviewers of the papers concerned. In the same way, Mann's claim that he did not delete emails was all very well, but readers of the report were left in the dark as to precisely what he did do when he received Jones' request to delete his Fourth Assessment Report correspondence.

The report revealed that shortly after this interview, Mann was asked by the inquiry team whether he still had the relevant email correspondence and he was apparently able to produce a selection of emails shortly afterwards, including some relating to the Fourth Assessment Report. One explanation of this is of course that Mann, like Jones and Briffa, had simply transferred all his emails to a memory stick and then deleted them from his inbox, thus allowing him to deny their existence while still being able to produce them for the inquiry. As far as we can tell from the inquiry report, however, details like this were never probed.

At no point in its work did the inquiry panel interview any of Mann's critics, but the report did reveal that they had interviewed two outsiders. Their choices are surprising and rather intriguing. Firstly there was Gerald North, an eminent atmospheric physicist, who was best known to sceptics as the chairman of the NAS panel into paleoclimate.* Although their report had confirmed McIntyre's major criticisms of the Hockey Stick, North and his team had tried to divert attention from the criticism by arguing that the Hockey Stick remained plausible because of its alleged similarity to other temperature reconstructions, quietly overlooking the fact that these relied on the same flawed data as the Hockey Stick – the bristlecones – that they had already condemned as unreliable. More disturbingly, North had spoken out in defence of the Hockey Team in an interview published just after Climategate. Asked about the trick to hide the decline, North told a journalist that he could see nothing wrong in what Jones had done.

Was data manipulated? I do not think so. In the NAS 2006 Report

---

*See p. 26.

on Reconstruction of Surface Temperatures for the last 2000 Years
...we constructed our own hockey stick curve. We put the tree
ring record on the graph and stuck the instrument record on for
the last 50 years in exactly the way [Phil] Jones in his [leaked]
email referred to as a 'trick'.[168]

Unfortunately, this was not true. The spaghetti graph in the NAS report did not include Briffa's series at all, so there was no need to delete
the divergence in the way Jones had done. And far from splicing two series together in the manner of the hide the decline graph, the NAS panel
had simply plotted the instrumental record alongside the proxy reconstructions.

With North having given such an erroneous version of events – a remarkable thing given how close he had been to the paleoclimate debate –
sceptics were very concerned at his involvement in the Penn State inquiry.
However, although it was not widely known at the time, the second scientist contacted by the Penn State inquiry also had some past involvement
in the Hockey Stick story. Donald Kennedy was the former editor-in-chief
of *Science,* and had at one time or another also been a senior official in
the Carter administration and President of Stanford University, a position from which he had resigned after a scandal involving the university
overbilling the Federal government. Attentive readers may have noted,
however, that Kennedy's name has already appeared in this story, when
he played a bit part in the efforts to discredit the Soon and Baliunas paper in the US senate.* Kennedy had also intervened at later stages of the
Hockey Stick story when he had objected to the US House of Representatives taking an interest in the affair.[169]

Unfortunately, these intriguing details were not known at the time
of the publication of the Penn State inquiry report, and most attention
was focused on North, and in particular comments he had made in the
aftermath of Climategate saying that he had not read the emails out of
'professional courtesy'.[170] It is unclear therefore how his advice could
have helped the inquiry, unless of course he had changed his attitude in
the meantime.

Given their backgrounds, it is perhaps unsurprising that North and
Kennedy were both, as the inquiry described them, 'very supportive of
Dr. Mann and of the credibility of his science', and their evidence seems
to have played an important part in the findings of the inquiry: on each

---

*See p. 10.

of the first three allegations the panel decided that there was no evidence to substantiate the allegations. Only on the fourth allegation – that his behaviour might have been inconsistent with accepted standards within the academy – were the panel less certain, and here they ruled that there was a case to answer and that this case should be passed to the full panel of investigation.

The inquiry's decision that there was no case to answer on most of the allegations bears careful examination. For example, on the charge of falsification, it was reported that the trick to hide the decline was an entirely innocent procedure:

> The so-called 'trick' was nothing more than a statistical method used to bring two or more different kinds of data sets together in a legitimate fashion by a technique that has been reviewed by a broad array of peers in the field. [167]

As we have seen, this is far from the truth, but as regards the inquiry the pertinent issue is not that the panel repeated obvious falsehoods, but that they had purported to investigate and clear Mann of the charge made. The panel of inquiry was, it should be remembered, equivalent to a grand jury, determining whether there were meaningful allegations to investigate. Its purpose was not to clear the accused – that much was clear from the university's procedures:

> *Inquiry* is defined as information-gathering and preliminary fact-finding to determine whether an allegation or apparent instance of research misconduct warrants an investigation.
>
> *Investigation* is defined as a formal examination and evaluation of relevant facts to determine whether research misconduct has taken place or, if research misconduct has already been confirmed, to assess its extent and consequences and determine appropriate action. [171]

That sections of data had been deleted and spliced was acknowledged by all parties. There was clearly a case to answer, therefore, and the question of whether what Mann had done in the Third Assessment Report was acceptable or not was a question for the Investigation panel. The inquiry had in essence usurped the role of the investigation.

The findings on email deletions were little more credible. The panel appeared to be satisfied with Mann's statement that he had not deleted emails and they were clearly also impressed by his ability to produce some emails related to the Fourth Assessment Report when requested. Precisely what he did when Jones asked him to delete emails was not asked, and so the charge as laid:

> Did you engage in, or participate in, directly or indirectly, any actions with the intent to delete, conceal or otherwise destroy emails, information and/or data, related to [the Fourth Assessment Report]... [167]

...had not in fact been answered. The possibilities of concealment and indirect involvement were simply not addressed.

Even on the question of Mann's conduct, where the panel members had not felt able to clear Mann without a hearing, the panel's report contained much that was readily shown to be false. For example, they claimed that:

> In 2006, similar questions were asked about Dr. Mann and these questions motivated the National Academy of Sciences to undertake an in depth investigation of his research. [167]

Given that the inquiry panel had interviewed North, the chairman of the NAS panel, it is extraordinary that they could have ended up with such a mistaken impression of the work it had performed – North's inquiry in fact had examined the *science* of paleoclimate, devoting only a fraction of its time to Mann's work and none at all to the question of whether his conduct had been inconsistent with academic norms.

If the inquiry's usurping of the remit of the investigations committee were not bad enough, they also appeared to have been playing extremely fast and loose with their own procedures. According to the university policy for such inquiries, the report was supposed to include any transcripts of interviews they might have made. [171] However, although the report acknowledged that transcripts had been made, they were nowhere to be seen in the report. Likewise, the policy required that the report reproduce the evidence examined in sufficient detail to allow readers to understand the decisions that had been made, but in the event the panel merely noted the types of information they had examined:

> ...e-mails, journal articles, OP-ED columns, newspaper and magazine articles, the National Academy of Sciences report ...and various blogs on the internet. [167]

While one of the allegations – of inappropriate conduct – would go forward to the Investigation committee, it was clear from the work on the inquiry panel that something was seriously amiss with the integrity of Penn State's examination of Mann's doings. Despite this, Penn State President Graham Spanier told journalists that the inquiry had 'taken the time and spent hundreds of hours studying documents and interviewing people and looking at issues from all sides',[172] a strange version of events given the failure of the panel to interview any of Mann's chief critics.

The university authorities were already beginning to look as if they were accomplished whitewashers and word of the problems started to spread. In particular, Republicans in the US House of Representatives began to make fierce criticisms of the university and there was a call for the Department of Justice to launch an independent investigation of Mann.[173] From conservatives in Pennsylvania there were calls for the state legislature to launch an investigation of the university. The end of the Penn State story was a long way off.

CHAPTER 7

THE SCIENCE AND TECHNOLOGY COMMITTEE

The House of Commons Science and Technology Committee is generally something of a sleepy backwater among parliamentary committees, with none of the headline-grabbing prominence of, say, the Home Affairs or Treasury select committees. However, with its attention now directed at the hotly contested area of global warming, and in particular the Hockey Stick wars, it was about to find itself in the international spotlight.

Scientists are few and far between on the benches of the House of Commons, but fortunately in 2010 a handful of well-qualified people were to be found among the members of the Science and Technology Committee (see Table 7.1). So while committee chairman Phil Willis was a history teacher by background, alongside him were several MPs with higher degrees and one full professor.* There were also two men who had been industrial chemists, and sceptics noted that one of these, Graham Stringer, had a voting record that suggested he was far from convinced by the evidence for manmade climate change. But Stringer seemed to be the exception – most other members of the committee appeared to be firm, or even fervent, supporters of climate change legislation, and there was little expectation of a favourable outcome. Fears were exacerbated when it was realised that many members of the committee were due to step down at the general election, only ten weeks away. So to add to concern that the committee members were mostly drawn from the opposite side of the global warming debate, there was now the worry that they might be 'demob happy' as well.

Nevertheless, interest in the hearings was high and there were nearly 100 submissions of evidence. Some of these were from those directly involved – UEA on the one side and, ranged against the university, their sceptic accusers, many of them familiar names – Steve McIntyre and his co-author, Ross McKitrick; Steven Mosher, who had first announced the Climategate disclosures; David Holland, who had made the fateful request for Briffa's emails; Doug Keenan, the sceptic who had publicly accused Jones of fraud. There were others too: Sonia Boehmer-

---

*MPS Desmond Turner and Bob Spink have PhDs in immunology and electronic engineering, respectively. Brian Iddon had been a professor of chemistry and Evan Harris was a medical doctor.

TABLE 7.1: Members of the Science and Technology Committee at the time of the first Climategate hearings

Name	Party
Phil Willis	Liberal Democrat(Chair)
Roberta Blackman-Woods	Labour
Tim Boswell	Conservative
Ian Cawsey	Labour
Nadine Dorries	Conservative
Evan Harris	Liberal Democrat
Brian Iddon	Labour
Gordon Marsden	Labour
Doug Naysmith	Labour
Bob Spink	Independent
Ian Stewart	Labour
Graham Stringer	Labour
Desmond Turner	Labour
Rob Wilson	Conservative

Source: STCR I.

Christiansen, the editor of the sceptic-friendly journal *Energy and Environment*, Benny Peiser, the director of GWPF; and Warwick Hughes, who had made that first request for CRU's data some five years earlier. Occupying the middle ground, there were submissions about the other inquiries – one direct from Muir Russell and his team, and another from Richard Thomas, a former Information Commissioner. There was also input from several learned societies – the Royal Society of Chemistry, the Institute of Physics and the Royal Statistical Society – together with submissions from many members of the public offering expert advice on particular aspects of the Climategate affair, such as the quality of the computer code produced at CRU.

It is extraordinary how little support there was for UEA in the submissions of evidence. Of the 57 individuals or organisations who wrote to the committee, only a handful could be said to be supportive of CRU – a climatologist called Peter Cox and a green pressure group, the Public Interest Research Centre, plus the submissions from UEA itself. One environmentalist who submitted evidence seemed shaken by what she had discovered about the quality of climate change science, which she described as 'a monster, whose tentacles appear to have gone very deep, with bad science in many areas'.[174] Even neutral submissions were few and far between; the vast majority were highly critical of Jones and his colleagues, a wave of denunciations of CRU and a litany of stories of declines hidden, journals threatened and emails deleted. If the committee were going to assess the Climategate affair on the weight of evidence alone, they were going to find it hard to come any conclusion other than that something was seriously amiss.

Much of the attention of outsiders was focused on the two key sets of evidence – McIntyre's submission and the one from UEA. We have already seen that the UEA evidence had been highly controversial, appearing to mislead the committee over the views of the Information Commissioner.* But there were many other aspects of the submission that attracted comment. For example it was noted that there was an odd structure to the document: it was described as being the submission of UEA, but certain sections were then flagged as having been written by CRU. It almost looked as if those at the top of the university heirarchy had wanted to distance themselves from Jones' responses.

In the days before the hearings took place, the commmittee released the names of the witnesses it was going to invite to give oral evidence and there was dismay among many of the sceptic onlookers. Despite being absolutely central to the Climategate affair, their names appearing again and again in the emails, neither McIntyre nor McKitrick had been invited to speak. Instead, the committee had invited Lord Lawson and Benny Peiser of the GWPF. Interestingly neither of these men would describe themselves as sceptics, with Peiser in particular maintaining that he is agnostic on the science. Either way, the sceptics with the most in-depth understanding of CRU's science and the background to the emails were to be shunned.

From the other side, Acton and Jones would be appearing on behalf of UEA and Muir Russell would appear to explain the work of his inquiry. Then there was the former Information Commissioner, Richard Thomas, and a panel representing the scientific establishment – Sir John Beddington,[†] the Met Office chief scientist Professor Julia Slingo, and a former head of the IPCC, Professor Bob Watson. These were, for the most part, familiar names to sceptics. Watson, in particular, had been ever-present in the media since the Climategate story first appeared, fighting a valiant

---

*See p. 112.
†See p. 93.

rearguard action in defence of the reputation of the CRU scientists. As we have seen, he had also been considered as a possible member of the Russell panel.* Meanwhile, Slingo had taken up a prominent role since the leaking of the emails, being pivotal in organising the public declaration of support for the integrity of climatology.† Beddington, a biologist by profession, was more of an unknown quantity, having had a much less prominent role in the climate debate and having only occupied his position for a few months, although, as we have seen, he had been working on the Climategate affair since soon after the story broke. The overwhelming impression was that the hearings were designed to close with a strong message of support for CRU.

The hearings took place on 1 March 2010 in the grand surroundings of one of the committee rooms of the House of Commons. Many of those who had submitted evidence were present in person, while others watched live video feeds or read live blogs of what was going on. David Holland and Jones' colleague Tim Osborn can be seen in the background of the video, furiously taking notes.

Lawson and Peiser were the first to be heard and it quickly became clear that several members of the committee intended to treat this part of the hearings as a trial of the two sceptics rather than an investigation of CRU and its staff. Even in the opening exchanges, Phil Willis managed to accuse Lawson of criticising the composition of the Russell panel, suggesting that this was because he was 'losing the arguments'. Soon afterwards, the panel started to question Lawson about GWPF's funding arrangements, the slightly surprising source of the first question on this subject being the man expected to be most supportive of the sceptic position, Graham Stringer. After the initial question was put there was a long series of follow-up questions on the same subject – Phil Willis, Stringer again, Evan Harris, then Willis again. This was a very strange way for the committee to conduct an inquiry into CRU: the overwhelming impression was that the committee hoped to cast doubt on the credibility of the GWPF men right from the start.

The questioning of Jones was something of a contrast. When he took his place in front of the committee there was shock at his changed appearance. Gone was the familiar confident scientist of the official UEA photograph, his place taken by an older, cowed figure with gaunt, be-

---

*See p. 103.
†See p. 99.

spectacled features and a dull expression, his hands visibly shaking. The weeks in the public eye had clearly taken their toll. In fact so worried were the university authorities that they had appointed a specialist public relations firm* and Jones had been given intensive coaching ahead of the hearing. It emerged afterwards that the committee had been told that he was close to suicide. This might explain why their questioning was very gentle – possibly too gentle, given the importance of the allegations against Jones. As we will see, at times the committee almost appeared to be steering him towards safety.

THE TRICK TO HIDE THE DECLINE

In the immediate aftermath of the emails' disclosure UEA had issued an apparently panic-stricken press release defending Jones' actions in hiding the decline.

> One particular, illegally obtained, email relates to the preparation of a figure for the WMO Statement on the Status of the Global Climate in 1999. This email referred to a 'trick' of adding recent instrumental data to the end of temperature reconstructions that were based on proxy data... To produce temperature series that were completely up-to-date (i.e. through to 1999) it was necessary to combine the temperature reconstructions with the instrumental record, because the temperature reconstructions ... ended many years earlier whereas the instrumental record is updated every month. The use of the word 'trick' was not intended to imply any deception. [109]

Unfortunately, this claim had a significant flaw. Despite the university's pronouncement to the contrary, Briffa's data series actually ran right through to 1994 – hardly 'many years earlier' – and, as we have seen, it had been truncated before being spliced to the instrumental data. The suggestion that hiding the decline had something to do with a lack of suitable proxy data was therefore demonstrably untrue. To make it worse, the university's press release had continued with a statement from Jones explaining exactly what had been done and thus making the deception plain:

---

*The Outside Organisation were later accused of having hacked the telephones of members of the public in an entirely unrelated scandal. Neil Wallis, the firm's chief executive, who was responsible for helping UEA, was arrested in connection with these allegations.

> One of the three temperature reconstructions... [in the WMO graph]
> shows a strong correlation with temperature from the 19th cen-
> tury through to the mid-20th century, but does not show a realistic
> trend of temperature after 1960. This is well known and is called
> the 'decline' or 'divergence'. The use of the term 'hiding the decline'
> was in an email written in haste. CRU has not sought to hide the
> decline... It is because of this trend in these tree-ring data, [which]
> we know does not represent temperature change, that I only show
> this series up to 1960 in the WMO statement. [109]

The preponderance of the written submissions to the Parliamentary
inquiry were critical of CRU, and the committee had no shortage of advice on the meaning of the trick to hide the decline, including of course
a definitive treatment in McIntyre's evidence. [175] This being the case, the
university was going to need a much more consistent and credible argument if they were to get Jones off the hook and they do not appear to
have wasted the weeks since the announcement of the hearings.

In fact much of the story advanced in the university's evidence was
factually in broad agreement with McIntyre's: after 1960, Briffa's series
diverged from the instrumental data and had therefore been shown to
be unreliable. The fact that the decline had been deleted and replaced
with instrumental data was also undisputed. As Jones explained in his
evidence:

> ... 'hiding the decline' referred to the method of combining the
> tree-ring evidence and instrumental temperatures, removing the
> post-1960 tree-ring data to avoid giving a false impression of de-
> clining temperatures. What it did not refer to was any decline in
> the actual thermometer evidence of recent warming. [176]

Jones noted that the divergence problem had been discussed by CRU
in the scientific literature and, as might have been expected, he also repeated the familiar narrative that the word 'trick' referred to 'a good way
of doing something'. This was, of course, misleading, because, as was
now well-known to sceptics, although the divergence problem had been
discussed in the literature, it had been hidden in reports for policymakers
– the IPCC assessment reports and the WMO report. Revealing uncertainties in the data in the expert literature but 'hiding' them in documents for
the use of politicians should have been seen as unacceptable by all of the
members of the committee yet, remarkably, when it came to the questioning of Lawson and Peiser, some of the committee members seemed very

keen to put a different spin on the episode. In the initial exchanges on the subject, Lawson told Phil Willis that he accepted that the word 'trick' could be used in the sense that Jones and his supporters were claiming he had meant it, but said that he took exception to the *hiding* of the decline. When Harris took over the questioning, he asked Lawson to reiterate this statement and went on to ask a question that was both factually inaccurate and extremely leading:

> Is it fair to say that... if the [Russell] panel decide... that 'hiding the decline'... was a legitimate way of treating the data to show, with an explanation, that a better way of looking at the data was not to allow the later tree ring data to influence the overall position – if they could show that that was argued in the publications; if the review panel comes to that conclusion – then you would be satisfied...?[177]

Harris's precise meaning is somewhat unclear, but later in the day he provided some clarification. During his questioning of Jones, Harris noted what he described as Lawson's 'assertion that when you [hid the decline] it was not set out in the publications', and contrasted this with the UEA statement that 'it' was 'part of the published record'. What Harris meant by 'it' in this sentence is obscure: he could have meant the existence of the divergence problem or alternatively its deletion. In the same way, what he meant by 'the publications' and 'the published record' was open to interpretation – he may or may not have included reports for policymakers, such as the WMO report and the IPCC assessment reports, in these terms. The distinctions are crucial to the question of the trick to hide the decline: the divergence problem *is* thoroughly discussed in the primary scientific literature, including in Briffa's own papers – this is undisputed – but the decline has also been deleted and sometimes replaced with instrumental data in documents written for politicians. Where this happens, the hiding of the decline is not generally explained and the reader remains unaware of what has happened. Harris's suggestion that the deletions had been explained is therefore largely untrue.*

After the hearings, when the committee came to deliberate their reaction to the trick to hide the decline, Stringer made an attempt to change

---

*An explanation of the deletion was included in the Fourth Assessment Report after a failure to do so in the drafts of the report led to protests by sceptics. Briffa, however, refused to show the series as it appeared in his own original paper, with the decline in place, but inserted some additional text discussing the problem after the final review had taken place.

the wording of the report, suggesting that they should state that they had not heard sufficient evidence to reach a firm conclusion.[178] Given the time constraints under which they were working, this might have appeared to be a prudent course of action and one around which the whole committee could unite. Remarkably, however, the three other committee members present for the vote – Harris, Boswell and Iddon – voted down Stringer's amendment. When the report appeared at the end of March 2010, it simply parroted the line that had been put forward by *RealClimate* three months earlier. 'Hiding the decline' was apparently 'shorthand for discarding data known to be erroneous', a choice of words that flew in the face of all the evidence the committee had heard, since even CRU had not suggested that there had been an error in the data and everyone agreed that the reason for the divergence was unknown.

The committee also suggested that by discussing the existence of a divergence in *Nature*, Jones was justified in *deleting* that divergence in reports for policymakers, since the *Nature* paper was cited in the report. The theory seemed to be that the politicians would read the report and refer to the primary scientific literature it cited, where they would discover the truth – namely that the graph they were seeing in the report should actually be a completely different shape. However, even in the unlikely event that the readers of the reports actually did this, they would have remained in the dark about how the transformation had taken place – there was no explanation of the steps involved in the various tricks to hide the decline in any of the documents written for policymakers.

These were extraordinary conclusions and appeared to represent an open invitation for scientists to mislead politicians in expert reports. But perhaps the most disturbing aspect of the committee's conclusions was that they had decided to ignore one of the best-informed witnesses completely: McIntyre's evidence on hide the decline was not even mentioned in the report.

PERVERTING PEER REVIEW

The question of the perversion of the peer review process had been raised in several submissions to the committee. We have already seen two alleged examples – the discussion of getting rid of von Storch after the publication of Soon and Baliunas's paper and the removal of James Saiers as the editor responsible for McIntyre's paper in GRL.* However, there were

---

*See p. 6 and p. 14.

several other examples of the undermining of peer review in the emails. Unfortunately, submissions to the committee were limited to 3000 words, and McIntyre had used much of his space discussing technical areas, such as the trick to hide the decline. He was therefore only able to devote a few lines to the subject of peer review, limiting himself to a few quotations from the Climategate emails:

> If published as is, this paper could really do some damage. It is also an ugly paper to review because it is rather mathematical, with a lot of [high level statistics] in it. It won't be easy to dismiss out of hand as the math appears to be correct theoretically.

> Recently rejected two papers (one for JGR and for GRL) from people saying CRU has it wrong over Siberia. Went to town in both reviews, hopefully successfully. If either appears I will be very surprised.

> I am really sorry but I have to nag about that review – confidentially I now need a hard and if required extensive case for rejecting.

> I can't see either of these papers being in the next IPCC report. Kevin [Trenberth] and I will keep them out somehow – even if we have to redefine what the peer-review literature is![179]

There was more to McIntyre's case than simply the accusation that Jones and his colleagues were trying to keep dissenting papers out of the literature. There was also a series of emails that suggested that the same group of scientists were giving 'soft' reviews to their colleagues, a process dubbed 'pal review' by one wag. However, with no more space to give quotations, McIntyre was reduced to citing examples of this kind of behaviour from Mann, Schmidt, Wahl and Ammann and another close colleague of the Team, Ben Santer.

The information he had been forced to omit would have presented a very disturbing picture of the way Jones reviewed the work of the other members of the Hockey Team. For example, the Climategate disclosures contained a review he had done on a paper by Michael Mann, in which the CRU man said:

> The paper is generally well written. I recommend acceptance subject to minor revisions. I will leave it to the editor to check that most of my comments have been responded to.[180]

Jones had also prepared a review of a paper by Gavin Schmidt – the climate modeller and *éminence grise* behind *RealClimate* whom we met in Chapter 3. Schmidt had written a critique of two sceptic papers on temperature trends in the troposphere – an important area of research because climate models projected that the troposphere would exhibit rapid warming – a so-called 'fingerprint' of manmade global warming. However, just as in his review of the Mann paper, Jones essentially rubber-stamped Schmidt's manuscript for publication, only requiring minor amendments to be made:

> This paper is timely as it clearly shows that the results claimed in [the sceptic papers] are almost certainly spurious. It is important that such papers get written and the obvious statistical errors highlighted... There is really no excuse for these sorts of mistakes to be made, that lead to erroneous claims about problems with the surface temperature record.
>
> My recommendation is that the paper be accepted subject to minor revisions. [181]

Also cited in McIntyre's evidence was Jones' review of the Wahl and Ammann paper, which, like the others, was overwhelmingly uncritical:

> This paper is to be thoroughly welcomed and is particularly timely with the next IPCC assessment coming along in 2007. The availability of the data and the programs on a website will go a long way to silencing the critics. I suspect though that this will not be the last word on the subject. [182]

Jones' written evidence to the committee only touched very briefly on the subject of peer review, mentioning just two of the specific allegations. One of these was the Soon and Baliunas affair, where he and his colleagues had discussed getting rid of von Storch as editor of *Climate Research*.* Jones' explanation was simply that the paper was a poor one, and he noted the subsequent resignations of members of the journal's editorial board. He did not offer an explanation of why trying to get rid of von Storch was a reasonable response to a paper he felt was poor, and nor did he mention the fact that he had met with von Storch to discuss the Soon and Baliunas affair in the weeks before the resignations from the journal's editorial board.

---

*See p. 6.

Jones also briefly discussing the incident in which he had said he and Trenberth would keep sceptic papers out of the Fourth Assessment Report. Here he said, he had only been 'expressing doubts about the scientific rigour of the two papers',[183] one by McKitrick and Michaels and the other by de Laat and Maurellis.* McKitrick is of course, Ross McKitrick, McIntyre's regular co-author.

The McKitrick and Michaels paper was an examination of the important CRUTEM surface temperature series, showing that there were correlations between changes in temperature and industrialisation in many areas. This was surprising because the global temperature data had been corrected for UHI and, as we have seen, Jones had claimed to show that the magnitude of the effect was small anyway.† If the correlations McKitrick and Michaels had found were real, it would imply that the UHI effect was much larger than had been thought, and that the rate of global warming was lower – on this measure at least. De Laat and Maurellis had reached broadly similar conclusions using a different methodology.

So when Jones said that he wanted to keep the two papers out of the IPCC reports, it is clear that he had a strong incentive for doing so and in fact there is evidence that he did attempt to put his words into practice. When the First Order Draft of the Fourth Assessment Report appeared, there was indeed no mention of either of the papers and sceptics among the reviewers were quick to protest. Despite this, in the Second Order Draft there was still no discussion of either paper and further protests ensued. In the final version of the report, Jones finally inserted a few lines about the McKitrick paper, but in order to negate any possible adverse effect he also said that the results in the paper were 'statistically insignificant' – in other words that they could not be relied upon. Because Jones' words were only inserted in the final draft of the report, they were not subject to any expert review and it is unlikely that they would in fact have survived any critical analysis, as McKitrick explained in his evidence to the Science and Technology Committee:

> [Jones'] claim that our results were statistically insignificant is inaccurate and was made without any supporting citation. To my knowledge no study showing such a thing exists. . .

*It is possible that Jones' message has become slightly garbled in this section, because he referred to Soon and Baliunas rather than de Laat and Maurellis. This does not appear to matter greatly to the case being made here.

†See p. 45.

[The statement] is unsupported, and in the context appears to reflect a fabricated conclusion.[184]

This allegation of fabrication is among the most serious that has emerged from the Climategate emails and surprisingly was entirely new to most of those who were following the hearings. It is remarkable to note that, even four months after the initial disclosure of the emails, new, serious and highly damaging allegations were still coming to light.

Cleverly, McKitrick also set out what the committee needed to prove or disprove his allegation: he explained that it was simply necessary for them to ask Jones to provide them with the scientific paper that supported his claim and, in particular, one that contained a *p*-value – a measure of the statistical significance – that would support Jones' claim. And, as McKitrick added, an inability to do so would be highly revelant to the committee's stated objective of assessing the implications of the Climategate emails for the integrity of scientific research.

Despite possible breaches of peer review being among the most serious allegations levelled at CRU, the committee chose to take oral evidence on the subject only from Jones. The questioning was led by Evan Harris, who unfortunately rather avoided the issue, addressing a number of incidents that were nothing to do with the allegations and which barely touched upon the subject of peer review at all.

Harris first asked about an incident involving the editor of *Energy and Environment*, Sonia Boehmer-Christiansen, who is emeritus professor of geography at the University of Hull. Boehmer-Christiansen had written to some UK civil servants to bring to their attention serious allegations made about CRU. Jones' response had been to write to her head of department insinuating that she should not be permitted to use her university affiliation, but he had been rebuffed.[185]

Harris's second line of inquiry concerned Jones' UHI paper, which was a very surprising decision. Keenan had sent the committee a detailed account of his fraud allegation against Wang, and in this he had made his allegations against Jones explicit – Jones' citation of the UHI paper in the IPCC's Fourth Assessment Report was, Keenan said, tantamount to scientific fraud because Jones knew that his findings were untenable. However, nowhere did the story of the paper and Keenan's allegations touch on the issues of the integrity of peer review or threats to journals (although Keenan documented many statements made by Jones that he said were false and misleading during the peer review process and also noted that Jones had breached peer review confidentiality on several occasions[186]). Quite how Harris arrived at the erroneous conclusion that it did remains unexplained, but the result of his mistake was that the committee ended up investigating a non-existent allegation.

So neither of the two incidents examined actually related to the specific allegations – of keeping sceptic views out of the literature using threats against the journals. It was a remarkable oversight by the committee, but perhaps not as surprising as another aspect of their investigation of the allegations relating to peer review: Jones was not at any point asked if he had actually made contact with any of the journals he had discussed threatening.

With barely a word put to any of the witnesses on the subject of peer review, the eventual contents of the report contained few surprises; the committee chose to focus entirely on the spurious examples raised by Harris. Inevitably, their conclusions were simply another echo of the story issuing from CRU itself:

> The evidence that we have seen does not suggest that Professor Jones was trying to subvert the peer review process. Academics should not be criticised for making informal comments on academic papers.[187]

Remarkably, no consideration was given to the question of 'pal review' either, and once again McIntyre's evidence was not mentioned or cited. Even more noteworthy was the failure of the committee to mention McKitrick's allegation of fabrication against Jones. They had simply turned a blind eye to the mountain of evidence against CRU's director and pronounced him exonerated. It was a remarkable performance.

THE INFORMATION WAS NOT FREE

The allegations of breaches of the FOI legislation were not among the areas selected by the committee for investigation – they were perhaps reluctant to do anything that might compromise the ICO's investigation despite it being clear that no criminal prosecution was going to take place. However, it was going to be impossible for the committee to ignore the FOI allegations completely, as they had been repeated in so many of the submissions of evidence. As we have seen, there was certainly a great deal of documentation to support the claims that these laws had been broken, possibly with the connivance of the university's FOI officer and even of the vice-chancellor himself.*

In the wake of the initial disclosure of the emails, UEA had adopted a conciliatory approach to their data in order to try to manage the storm of media criticism that was enveloping them. At the end of November 2009, they had announced that they would be releasing all of the raw data in due course, but that first they had to obtain permission from all the national meteorological services whose permission was said to be required. However, they went on to say that the vast majority of the data was actually already available through the international databases such as GHCN. As we have seen, this argument is somewhat misleading because without knowing the precise identity of the stations used (and because of the errors in the list released by CRU in 2007, these were still not known with certainty) it was impossible to get hold of the right data. Even then, a copy of the data from GHCN was not actually what the sceptics were after, as Mosher explained in his written evidence to the select committee:

> There is a fundamental difficulty facing anyone who wants to vali-
> date the claim that [the CRU, NASA and NOAA series] use largely the
> same data. In order to validate this claim with certainty, one must
> have access to the data as used. That is, one must have the various
> copies each of the agencies used in constructing their series. The
> point is a fine one, but [this] is the typical first test in any quality
> assurance test. CRU claim to copy some of their data from GHCN.
> To verify this claim one must compare the data that GHCN has with
> the copy that CRU claim to have made. [109†]

So in essence there were two different versions of the data being discussed – the raw data as used by CRU and the copies of the raw data in the archives – and UEA repeatedly referred to the latter, while presumably being fully aware that sceptics wanted the former. For example, Jones claimed that a statement UEA had issued the previous summer had been misunderstood. In this statement it had been claimed that:

> Data storage availability in the 1980s meant that we were not able
> to keep the multiple sources [of data] for some sites, only the sta-
> tion series after adjustment [were kept]. [79]

---

*See p. 53.

†The issue might appear somewhat pedantic, but this is not so. Important examples of differences between the data as used and purportedly identical copies in climate
archives are discussed in *The Hockey Stick Illusion*. [188]

Now, months later, Jones was telling the committee that the data had *not* been lost.[189] However, this apparent contradiction had a simple explanation. The original statement had been talking about CRU's copy of the data, while Jones was now referring to the copies of the data in the archives. In essence he was implying that the raw data as used *had* been lost, but was saying that a copy – which presumably should be identical – was available elsewhere.

An exchange between Lawson and Peiser and Ian Stewart, one of the committee members, strongly suggests that the committee did not understand these subtle distinctions. Stewart observed that UEA had claimed that all the data was available, and was corrected by Peiser, who pointed out that they had only promised to release the data once permission had been obtained.* Stewart therefore appears to have been talking about the archive copies of the data while Peiser was presumably referring to the data as used, which is what the sceptics wanted to see.

The confusion over different versions of scientific data emerged again in Jones' evidence when he came to defend his colleague Keith Briffa's failure to release the Yamal tree ring data.† Jones' excuse was that Briffa had referred McIntyre to the Russian scientists who originally collected the raw data, so he was once again sidestepping the fact that McIntyre wanted the data as used, precisely so that he could check it against the original Russian source.[189]

Some of the questioning of Jones was now quite firm, particularly from Stringer. Jones admitted that he had resisted requests for data, mentioning in particular his notorious refusal of Warwick Hughes.‡ Jones admitted that he had 'written some very awful emails...',[190] but many of his responses to the committee were still misleading. When he said that the list of stations used in his temperature series had been published in 2007, Stringer asked how long it had taken to extract the information from him. Jones replied that it had taken 'about six months from the first FOI request', which was not only an admission of a breach of the FOI Act,§ but also failed to mention that Hughes had actually made his request in 2005. There was shock among the committee when, in the next exchange, Jones said that it had not been standard practice to make data and codes available to other scientists.

---

*See p. 53.
†See p. 55.
‡See p. 45.
§The Act requires public authorities to respond within twenty working days.

Either way, the questioning had little or nothing to do with what had actually happened at CRU, adding to the impression that the select committee was not keen to get to the bottom of things. This feeling of awkward questions being swept under the carpet was reinforced by the questioning of Richard Thomas, the former Information Commissioner. When Thomas was asked about events at CRU, he admitted that he knew none of the details. Presumably the current incumbent would also have had difficulty in responding to these kinds of questions, since he would have been unable to give details about an ongoing investigation. Because of these considerations, the questioning was of necessity limited to general issues such as whether the 60-day of limitations could be disapplied, the interaction of the FOI Act with other legislation in the area, and the ICO's squabble with UEA.*

The committee's report was relatively critical of UEA's handling of FOI requests, although given that Jones had effectively admitted to having breached the legislation, the mild criticism in the report might have been seen as rather inappropriate. So while the committee agreed that there was prima facie evidence that FOI legislation had been breached, they refused to mention any names, excusing themselves in the following terms:

> It would, however, be premature, without a thorough investigation affording each party the opportunity to make representations, to conclude that UEA was in breach of the Act. In our view, it is unsatisfactory to leave the matter unresolved simply because of the operation of the six-month time limit on the initiation of prosecutions. Much of the reputation of CRU hangs on the issue. We conclude that the matter needs to be resolved conclusively – either by the [Russell panel] or by the Information Commissioner. [187]

The suggestion that the ICO might complete its investigation into UEA was remarkable, since the Information Commissioner's staff had already made it clear that they were 'unable to take the matter forward'.† However, with nobody involved in the ICO's investigation being questioned, there was no opportunity for this oversight to be flagged up to the committee.

The committee closed by noting their 'concern' about UEA's handling of information requests and suggesting that UEA needed to review its

---

*See p. 113.
†See p. 113.

policy in this area. But no names were mentioned and no guilt assigned. It appeared that in this area too, nobody was to blame.

By the time the Science and Technology Committee met, the revelations about the team that Russell had chosen for his inquiry had already hit the headlines – Campbell had resigned and it was clear that Boulton's position was equally inappropriate. So when the questioning of Lawson and Peiser began, the composition of the inquiry panel was the first question raised by chairman, Phil Willis. However, when Lawson raised the subject of Boulton's continued presence on the panel he was diverted onto other matters by Willis:

> Some would say, Lord Lawson, it is because you are perhaps losing the argument here in terms of the panel that you are criticising them, is that fair? [191]

It was clearly going to be impossible to avoid addressing the independence and neutrality of the Russell panel entirely, but when the time came to question Russell himself, the conversation was quickly steered away from any controversy by Harris, who asked:

> The composition of your team has been criticised by people who can be described as coming from the climate sceptic point of view. Is it your ambition to satisfy them or do you recognise you may never satisfy some critics from that quarter? What is your outlook on that? [192]

The questioning of Russell continued in similar vein, with no attempt to determine how it was that Russell had chosen two panel members quite as inappropriate as Campbell or Boulton, or indeed to probe why it was that Boulton remained on the panel at all. One of the committee members, Tim Boswell,* chose to ask Russell whether his inquiry was mere 'window-dressing', and was unsurprisingly told that it was not – or at least that Russell hoped so. [193] Boswell's questioning then moved on to other areas such as the need for the inquiry hearings to take place in public, but he appeared to be satisfied with an assurance from Russell that notes of any interviews would be posted on the internet.

---

*By a strange coincidence, Boswell was David Holland's MP.

Even then, there was still another opportunity for the committee to help ensure that the Russell inquiry was fair and above board. When they came to deliberate the final report, Stringer proposed an amendment that would have ensured that Boulton, the activist–scientist, was balanced by a scientist with sceptic views. This would ensure that everyone on the panel was 'kept honest' and would prevent any suspicion of a whitewash. Once again, however, Harris, Boswell and Iddon voted him down.

THE REPORT

It was clear from the hearings that the committee was not intending to perform a diligent investigation into what had gone on at CRU and there was little doubt among sceptics that a full exoneration would be issued in due course. The first details of the committee's findings appeared on 30 March, on the eve of the report's publication, and there was little surprise at the contents – the whole inquiry appeared to have barely risen above the level of a sham. Indeed, one prominent observer, the sociologist Frank Furedi, thought it was just that:

> As [committee member] Doug Naysmith indicated, he hoped that the report would serve as a 'corrective' to climate-sceptic hysteria. Investigations that are meant to serve as a 'corrective' to people's misguided or immoral sentiments used to be called rituals. And that is what this the House of Commons Science and Technology Committee's 'limited inquiry' was mostly about: a ritualised pseudoinvestigation aimed at correcting people's allegedly backward views. [194]

In his press release, Willis had gone out of his way to avoid the specific allegations made as a result of the Climategate disclosures, briefly commenting on the trick to hide the decline and freedom of information, but returning repeatedly to the overall global warming hypothesis, which he declared undamaged by the CRU affair. Nevertheless, he looked forward to the report of the scientific assessment panel, which he said would determine whether the work of CRU was 'soundly built'. [195]

In a radio interview at the same time, Willis went even further, declaring that CRU's science was 'beyond reproach' and declaring that Jones was exonerated of the accusation of manipulating data because of the similarity of CRUTEM to other temperature indices, a position that rather avoided the fact that many of the principal accusations were of manipulating paleoclimate data rather than thermometer records and that Keenan's fraud allegation had all but been ignored.[196] But above all, Willis's comments left observers with a reinforced impression that his intention all along had been to protect the global warming movement rather than to find the truth.

Shortly after the hearings had ended there was a surprising development regarding some of the things Acton had told the committee about data availability. In the immediate aftermath of the emails' disclosure, part of UEA's crisis management strategy had been to tell the public that they were belatedly addressing the question of the availability of the weather station data that sceptics had been seeking for so long. As the university explained in its first public statement after Climategate:

> The raw climate data which has been requested belongs to meteorological services around the globe and restrictions are in place which means that we are not in a position to release them. We are asking each service for their consent for their data to be published in future.[108]

As we have seen, it is doubtful whether there were in fact any meaningful restrictions in place, but nevertheless the university's statement did at least represent progress. A week later Jones wrote to each of the national weather services asking for permission to publish data. However, as he now explained, there was a complication:

> We stress that the data we hold has arisen from multiple sources, and has been recovered over the last 30 years. Subsequent quality control and homogenisation of these data have been carried out. It is therefore highly likely that the version we hold and are requesting permission to distribute will differ from your own current holdings.[197]

The numbers that Jones was requesting were therefore rather different to what sceptics were actually seeking: Jones was asking for permission to release homogenised and quality controlled data, while the sceptics wanted the raw data. In fact, the semi-processed figures that Jones had asked for were CRU's property anyway. And at least one meteorological service – the Swedish Meteorological and Hydrological Institute (SMHI) – was not happy with what Jones was proposing. It looked to them as if Jones was proposing that CRU's data be passed off as Swedish.

Given ... that the version of the data from the SMHI stations that you hold are likely to differ from the data we hold, SMHI do not want the data to be released on your web site.[198]

Instead, they suggested, users might be referred to SMHI's own web-site.*

Remarkably, when the Science and Technology Committee had asked Jones and Acton about the availability of the data, Acton had given a reply that was highly misleading:

Unfortunately, several of these countries impose conditions and say you are not allowed to pass it on, so there has just been an attempt to get these answers. Seven countries have said 'No, you cannot', half the countries have not yet answered, Canada and Poland are amongst those who have said, 'No you cannot publish it' and also Sweden.[199]

Acton, however, was being watched closely around the globe and his comments were noticed by a Swedish sceptic group called the Stockholm Initiative who quickly realised that what Acton had told the committee was false. Shortly afterwards they requested copies of the relevant correspondence from SMHI under FOI.

The request to SMHI was a problem – Sweden has a strong culture of openness in government and Acton's suggestion that they were withholding their climate data was therefore very damaging to SMHI's reputation, as they complained in a letter to Jones shortly afterwards:

[Our earlier letter has led to] bad publicity both to SMHI and to the climate research community... Our response was based on your information that it was likely that the version held by you would... most likely differ from our current holdings. It has never been our intention to withhold any data but we feel that it is paramount that data that has undergone, for instance, homogenisation by anyone other than SMHI is not presented as SMHI data. We see no problem with publication of the data set together with a reference stating that the data included in the dataset is based on observations made by SMHI but it has undergone processing made by your research unit. We would also prefer a link to SMHI or to our web site where the original data can be obtained.[200]

---

*Note, however, that the website apparently only contained data over shorter timescales, a situation that SMHI indicated would be resolved in due course.

The following day all of the correspondence between Jones and the Swedes was published on the sceptic blogs, along with a press release from the Stockholm Initiative, who condemned the sleight of hand in no uncertain terms:

> *Climate scientist delivers false statement in parliament enquiry*
>
> ... [Jones'] statement is false and misleading in regards to the Swedish data. All Swedish climate data are available in the public domain. As is demonstrated in... correspondence between SMHI... and Dr. Jones... this has been clearly explained to Dr. Jones. What is also clear is that SMHI is reluctant to be connected to data that has undergone 'processing' by the East Anglia research unit. [201]

With the full story on public view, Acton had to back down and shortly afterwards a letter was dispatched from the university to the Science and Technology Committee to tell them that the situation with the Swedish data had changed, although sidestepping the question of whose data it was that was being considered for publication:

> [In his oral evidence]... Professor Acton outlined that a number of countries including Sweden had not given permission for UEA to publish data from their Meteorological Services on the UEA website. The information relating to Sweden was based upon a letter from the Swedish Meteorological and Hydrological Institute (SMHI) to Prof Phil Jones dated 21 December 2009. A second letter from SMHI received 8 March 2010 now gives permission for CRU to publish its Swedish data on the UEA website. [202]

CHAPTER 8

THE TRICK TO HIDE THE TRICK

At the start of February 2010, the launch of the Russell panel was imminent and the world was awaiting the Science and Technology Committee hearings, just a few weeks away. In the background, however, senior figures in the scientific bureaucracy were worried. On 6 February, Professor Alan Thorpe, the head of the National Environmental Research Council, wrote to Sir John Beddington to relay his concerns, suggesting that a separate inquiry be launched into CRU's science to determine whether it was sound or not.[203] Thorpe's suggestion seems to have been favourably received and further discussions ensued: the following day Acton emailed Russell to explain that he had been conferring with 'senior science figures' who were worried about how the parliamentary inquiry would view the fact that the Russell panel was not going to reassess the UEA's science. In order to pre-empt any criticism, Acton said, he intended to ask the Royal Society to convene 'an audit of CRU's key scientific publications, an external appraisal of the science itself'.[204]

Russell was in full agreement with Acton and it was agreed that the university would approach the Royal Society as soon as possible. Shortly afterwards Davies emailed Beddington to tell him the news:

> Muir Russell is launching [his inquiry] tomorrow. Our understanding is that he will state definitively that he will not be reviewing the 'science'. Given the time which has elapsed since we instigated the review. . . and other events, we are of the view that there should be as rapid as possible scientific assessment of key CRU publications. There has been discussion between the Royal [Society] (Martin Rees, Brian Hoskins), UEA (me, Peter Liss) and Alan Thorpe [National Environmental Research Council]. Initially we were hoping that the Royal [Society] would undertake this, but Martin feels it more appropriate that [they help us] us identify people with the appropriate standing, independence etc. We plan on issuing a statement to this effect tomorrow, Muir Russell has agreed.
>
> It is difficult to say anything about the time-scale until the assessors have been appointed but we want this to be done as quickly as possible. Will keep you in the loop and seek your advice.[205]

Beddington responded the next day, suggesting that the emphasis should be placed on the science and also saying that he would think of a suitable person to chair the new inquiry. Shortly afterwards, he came up with a suggestion in the shape of Ron Oxburgh – Baron Oxburgh of Liverpool – a former colleague from the days when they had both been senior figures at Imperial College, London.* He also suggested another former colleague from Imperial – the statistician Professor David Hand – as a panel member.

Oxburgh is a geologist and a former chairman of the UK arm of Royal Dutch Shell. On the face of it, therefore, he was admirably qualified for the role, but even a cursory examination of his background left him wide open to criticism: he had a large number of financial interests in green businesses, including chairmanship of two renewable energy companies and roles as an environmental adviser to companies such as Climate Change Capital, the Low Carbon Initiative, Evo-Electric, Fujitsu and Deutsche Bank. He was also a director of an organisation called Global Legislators Organisation for a Balanced Environment (GLOBE), which works to promote the introduction of green legislation around the world.

Oxburgh was well aware of the likely impact of his conflicts of interest and he was initially reluctant to take part. However, Beddington prevailed upon his friend to change his mind and by the end of February he was safely installed as chairman,[207] leaving only the problem of appointing the remaining panel members. Shortly afterwards, Davies sent an email to Martin Rees – Baron Rees of Ludlow – the eminent astrophysicist who was at the time the president of the Royal Society.

> Ron [Oxburgh] and we [have] settled on a list of 13 possible candidates for [Scientific Assessment Panel] membership. The target will be ~6 but we will need to have a larger choice to account for non-availability.[208]

Davies went on to list the 13 candidates and then, intriguingly, gave his views on their reliability on the question of climate change:

> We see Kelly, Huppert and Hand as being mainly neutral with respect to recent climate change (some neutral at most)... Out of

---

*No written record of this suggestion has emerged, but in a later email Davies is seen to thank Beddington for putting Oxburgh's name forward.[206]

these 13, we would hope to get 6 with a suitable range of ex-
pertises, and a range of 'attitudes' towards recent warming [and]
greenhouse gases – from those who already see it as a problem, but
without being right in the middle of the climate science community,
to those [who] will come to it with a questioning objectivity. [208]

Remarkably then, Davies appears to have been suggesting to Rees
that only some of the panellists would be expected to look at CRU's published work 'with questioning objectivity', and this point needs to be
borne in mind when considering the eventual list of people chosen. It
is also worth noting Oxburgh's comments in an interview shortly after
the eventual publication of the report in which he told the interviewer
that the scientists appointed were not asked for their views on climate
change but were appointed for their scientific expertise. [209]

Davies finished his email by asking Rees to review the list for suitabil-
ity, suggesting that he should also involve 'Brian' in this process. Brian
is Sir Brian Hoskins, who we have already met as one of the IPCC review
editors who had received one of David Holland's FOI requests. He had
also been involved in the early discussions about setting up a review of
the science.*

Davies' request for approval of the shortlist strongly suggests that this
was Rees' first involvement in the selection process and that the names
on the shortlist of candidates had been selected by UEA itself. This view
is reinforced by the response to a later FOI request, in which Beddington
suggested that the identification of candidates had been a role primarily
undertaken by UEA, although he also said that the Royal Society had been
involved:

> The appointment process and selection conducted by UEA was in-
> formed by advice from the Royal Society, to ensure appropriate
> rigour, expertise and objectivity. As part of proper practice, in
> putting together a high quality panel the UEA leadership also took
> soundings on potential members, including candidates for the role
> of chair, from senior figures in the scientific community. [207]

Fourteen minutes after receiving Davies' email, Rees responded posi-
tively, saying that he thought it seemed a 'strong' list, although he emphasised his lack of expertise in the relevant science. His reply was copied to

---

*See p. 37 and p. 147.

Hoskins, along with Davies' original email with the shortlist and the discussion of the candidates' reliability. Shortly afterwards Hoskins emailed to say that he too was satisfied with the UEA list.

The following week, Oxburgh and Davies set about whittling down the 13 names that Rees and Hoskins had approved to their six preferred candidates and shortly afterwards Davies reported their decision to Beddington:

> Michael Kelly FRS – engineer, Cambridge
>
> Herbert Huppert FRS – mathematician, Cambridge
>
> David Hand FBA – Imperial
>
> Kerry Emanuel Mem.NAS – meteorologist, MIT
>
> Huw Davies – meteorologist, ETH Zurich
>
> Lisa Graumlich – tree-ring analyst, University of Arizona[210]

As we have seen, the first three names on the list were the candidates who were expected to look at CRU's work 'with questioning objectivity', while the others were considered to be convinced global warming believers: with Oxburgh in the chair, there would therefore be only be a minority who might be expected to address their task in an even-handed manner.

It was one thing to invite people to sit on the panel, but there was always a risk that some of them would not be willing to take part. Davies and Oxburgh, whose profiles in the academic firmament were not high, particularly in the USA, were going to need some assistance to ensure that all of the chosen candidates would agree to take part. As Davies explained to Beddington, it was therefore going to be necessary for some 'warming up' of the candidates to take place and this was already in hand, with Rees, Hoskins and Oxburgh all involved. There are hints, however, that 'warming up' may have meant slightly more than encouraging participation in the panel. At around the same time, Beddington's private secretary wrote to Davies to tell him that Beddington had followed up on the request to speak to Hand:

> John [Beddington] wanted you to know he has spoken to David Hand, who was in agreement with John's suggestions (and therefore has been 'warmed up').[211]

The equating of agreement with Beddington with being 'warmed up' is striking.

By the following week, with the panel in place, attention turned to the list of publications that would be examined. When he had advised them of the names of those selected as panel members, Davies had also circulated a list of papers to Rees, Hoskins and Beddington, and this had been circulated to the panel members. However, it was clear to the UEA team that as soon as the panel was announced, there would be intense pressure to reveal the list of papers that would be looked at and this was something that they did not want to do. As Davies told Rees and Hoskins,

> Initially we did not wish to do this but we have now been persuaded this is probably a good idea and it may, indeed, deflect other disruptive efforts by some in the media [and] blogosphere. Ron [Oxburgh] is comfortable with this, but is keen that we can say that it was constructed in consultation with the Royal Society.

> I did send you this list earlier, which I attach again here. They represent the core body of CRU work around which most of the assertions have been flying. They are also the publications which featured heavily in our submission to the Parliamentary Inquiry, and in our answers to the Muir Russell Review's questions.

> I would be very grateful if you would be prepared to allow us to use a form of words along the lines: 'the publications were chosen in consultation with The Royal Society'.[212]

There appears to have been no question of Rees or anyone else at the Royal Society actually putting forward papers for the review. These had already been chosen by UEA and sent out to the panel members. The Royal Society's role was simply to approve the list and to lend its name in order to give the appearance of independence and rigour.

Within minutes of receiving Davies' email, Rees advised that he had no objections to allowing the form of words suggested, provided that Hoskins was happy with the list, although he emphasised again that he had no expertise in the area himself. Hoskins, however, is a climate modeller rather than an expert in paleoclimate reconstructions or surface temperature measurements and he, like Rees, had to explain to Davies that he was not really qualified to comment, although he did not apparently see this as a problem.

> I am not aware of all the papers that could be included in the list, but I do think that these papers do cover the issues of major concern.[212]

As we will see, the position of Davies and Hoskins that the papers selected covered the most contentious issues was to be hotly disputed, and Davies, as a former head of CRU might well have been expected to know the true details.

Arrangements were hurriedly made for members to meet at UEA at the start of April. There was a minor hiccup when it was discovered that Graumlich was unable to attend at the same time as the rest of the panel, so a somewhat unsatisfactory arrangement was made whereby there would be two visits to UEA, with Graumlich spending two days on site at the end of March and the rest of the team attending the following week. Oxburgh and Hand would, however, accompany both visits.

With all the arrangements for the inquiry in place, there was now only the small matter of the public announcement of the panel members, but this was the cause of considerable concern. Everyone involved had witnessed the sorry sight of Philip Campbell being forced to resign from the Russell inquiry just a month earlier, and it was vital that nothing similar should happen to Oxburgh's team. A warning was therefore sent to all members of the panel asking that when they submitted their biographies for the press release they should make doubly certain that all links with UEA and the IPCC were disclosed.

The panel was announced on 22 March 2010 and the team seem to have decided that discretion was the better part of valour, as the list of papers to be examined was not mentioned at all.[146] However, there was no hiding the panel members' identities and, as the UEA team had expected, there was intense scrutiny of everyone involved. The press release claimed that Oxburgh had been appointed 'on the recommendation of the Royal Society',[146] a statement that was presumably intended to make criticism of the occupant much harder, but suffered from the drawback that it was untrue – as we have seen, Oxburgh's name had actually been put forward by Beddington.

As the watching sceptics examined the names of the panel members, Oxburgh's conflicts of interest were quickly noticed and Davies was forced to issue a statement defending the appointment. He said that, no matter what Oxburgh's interests were – and he added that the university had been aware of them all along – he was quite sure that the panel would be led 'in an utterly objective way'.[213]

Even worse, it was also discovered that Kerry Emanuel had made comments on the Climategate emails that appeared to prejudge the issues in exactly the same way as Philip Campbell's unfortunate radio interview had done. Speaking at a debate at MIT, Emanuel had said:

> What we have here...are thousands of emails collectively show-
> ing scientists hard at work, trying to figure out the meaning of evi-
> dence that confronts them. Among a few messages, there are a few
> lines showing the human failings of a few scientists...scientifically,
> it means nothing...[214]

Whether he was right or wrong on this, his words were a gift to the
sceptics, who also observed that many of Emanuel's other remarks in
the debate had been in a similar vein: he had spoken of a well-funded
public relations campaign that was attempting to drown out or distort the
message of climate science, linking this campaign to a corporate machine
that was 'branding climate scientists as a bunch of sandal-wearing, fruit-
juice drinking leftist radicals engaged in a massive conspiracy to return
us to agrarian society'.[214]

In the hours after the announcement of the panel, there were more
and more revelations. Emanuel had co-authored a paper with Michael
Mann, while Graumlich worked alongside one of the other Hockey Stick
authors, Malcolm Hughes, and the two panel members had contributed a
paper to a volume edited by Phil Jones. There were some futile interven-
tions from green journalists, which added little to the debate: the *Financial Times*'s Fiona Harvey wondered if sceptics would only be happy if a
sympathiser, such as Lord Lawson, were appointed to lead the inquiry,[215]
a suggestion that was greeted with contempt on the sceptic blogs. Meanwhile, Bob Ward, the chief PR man for the Hockey Team in the UK, tried to
negate some of the criticism before it happened by predicting that there
would be 'attempts by so-called "sceptics" to discredit the panel before it
has even started work'.[216] McIntyre, meanwhile, simply stated the way
so many felt:

> I think that the University of East Anglia had both an opportunity
> and an obligation to establish impartial and untainted inquiries.
> Kerry Emanuel is not an appropriate selection. Nor is Lisa Graum-
> lich. Nor is Lord Oxburgh. They should resign and let the balance
> of the panel proceed without them.[217]

THE INVESTIGATION

In due course, the scientists set to work. We know little of the impressions they formed as they worked, with one exception. Michael Kelly

had volunteered to look at the Briffa papers – a series of six studies dating back to 1998. These were all tree ring studies but, significantly, did not include Briffa's important multiproxy paper with Tim Osborn, which had been prominently cited in the IPCC's Fourth Assessment Report and over which serious questions of cherrypicking remained to be answered. Within a week of setting to work, Kelly had finished reviewing Briffa's papers and set down his impressions in a paper, which he then sent to Oxburgh via Acton's office – remarkably for an inquiry that was alleged to be independent of the university, all of the Oxburgh panel's correspondence was routed in this way. The following day, the paper was forwarded to Hand and Graumlich, who would be looking at the same Briffa studies on which Kelly had commented.

The Kelly paper gives a remarkable insight into the thinking of one of the panel members and is worth considering in detail. From reading Kelly's notes it is possible to see someone who is bending over backwards to give the benefit of the doubt to the scientists he is investigating – indeed he expressed full confidence in Briffa's integrity – but who has serious misgivings about what he is seeing:

> There is no evidence, as far as I am concerned, of anything other than a straightforward scientific exercise within the confines described above. The papers are full of suitable qualifications about the limitations of the data and the strength of the inferences to be drawn from them. I find no evidence of blatant malpractice. That is not to say that... choices of data and analysis approach might be made in order to strain to get more out of the data than a dispassionate analysis might permit.[218]

Several of the Briffa papers related to the subject of the divergence problem. As we have seen already,* it was undisputed that Briffa had addressed the divergence problem in the scientific literature; the sceptics' complaint was that CRU scientists had hidden the uncertainty created by the divergence problem when they wrote about it in the official reports for policymakers. This point was not lost on Kelly: having noted how hard it was to correlate the divergence with the theory of manmade global warming, he went on to express his concerns over the way the issue had been dealt with by the IPCC:

> Up to and throughout this exercise, I have remained puzzled how the real humility of the scientists in this area... is morphed into

*See p. 130.

statements of confidence at the 95% level for public consumption through the IPCC process. This does not happen in other subjects of equal importance to humanity. . . I can only think it is the 'authority' appropriated by the IPCC itself that is the root cause. [218]

Kelly was apparently deeply concerned that panel members would be on the receiving end of some harsh criticism if they were to issue an exoneration that was not extremely tightly defined:

> If we give a clean bill of health to what we regard as sound science without qualifying that very narrowly, we will be on the receiving end of justifiable criticism for exonerating what many people see as indefensible behaviour. [218]

He went on to note that several scientists had said that they saw 'prima facie evidence of unprofessional activity' in the emails.

Kelly's paper was quite out of the blue and was clearly going to be the cause of great difficulties if the intention was to whitewash the inquiry. Intriguingly, it appears that as well as being sent to Hand and Graumlich, Kelly's paper, or perhaps just word of its existence, may have found its way to Beddington: within 24 hours of Kelly sending his paper to Oxburgh, Beddington emailed Davies to tell him that he had spoken to Kelly:

> You may know that I also talked to Michael Kelly who was very positive and understood the absolute need for objectivity, particularly given his known stance. [219]

Little is known of the separate visit to UEA by Oxburgh, Hand and Graumlich at the end of March. The three panel members apparently met Briffa for a meeting that had been scheduled to last for an hour and a half, and then discussed in private what they had heard. A similar pattern was followed for the main visit. On the first day, the panel had an hour-long session with 'Phil Jones, Tim Osborn and team' – a 30-minute presentation from Jones followed by another 30 minutes of questions. It is not clear whether or not Briffa was in attendance. The panel then held private discussions for the rest of the day, returning to CRU for another hour and a half with 'Phil Jones, Tim Osborn and team'. The panel then appear to have spent the remainder of their time at UEA preparing the report, which was complete by the time they departed at 3:30pm on the second day. Extraordinarily, the Oxburgh inquiry's visit to the university consisted of a total of only four hours of meetings with the CRU scientists.

Just over three weeks after the announcement of the Oxburgh panel, rumours started to circulate in the blogosphere that the publication of the report was imminent. The news was greeted with a mixture of incredulity and scorn. It looked as if Oxburgh and his team felt it was only necessary to go through the motions of performing a diligent review and that they could then simply brush off any criticism. Within hours the gossip proved correct and the report was published on the UEA website, with a press conference at the Science Media Centre for invited journalists. The Science Media Centre also helpfully issued a crib sheet for journalists containing reactions from several prominent figures working in the area: by a remarkable coincidence, the people invited to comment were all names that we have encountered earlier in this story: Rees, Hoskins, Myles Allen and Bob Ward.[220]

The report was extraordinarily brief, filling just five sheets of paper when printed out in full, including all the references and the biographies of the panel members. However, in another remarkable coincidence, there was a consensus among the Science Media Centre's talking heads that this was not a problem: Ward described it as 'rigorous' and 'thorough', Hoskins said it was 'thorough and fair', and Rees opined that it was 'thorough' and 'authoritative'.[220] Judith Curry later gave a more credible assessment of Oxburgh's work, explaining that when she saw the report, she thought she was reading the executive summary and only later discovered that there was nothing else. The report, she said, had 'little credibility'.[221]

It was not just the brevity of the report that surprised readers, the content was shocking too. The first bombshell was the realisation that Oxburgh and his team had only looked at published papers and had failed to examine CRU scientists' work on the IPCC reports, where issues like 'hide the decline' and McKitrick's allegation of fabrication against Jones were among the most serious to emerge from the emails. Despite the earlier concerns of UEA staff, nobody among the sceptic community appears to have noticed this dramatic restriction on the panel's work at the time it was announced. As McIntyre noted, the only mention of the trick to hide the decline in the report was a veiled allusion towards the end, regretting the fact that 'the IPCC and others' had sometimes 'neglected to highlight' the divergence problem. It was hard for anyone watching to credit the idea that Oxburgh and his team were unaware that it was Jones and

Briffa who were at the centre of these most serious allegations. It was generally assumed that this statement represented a wilful and probably cynical turning of a blind eye. As McIntyre put it:

> ... the Oxburgh report is a feeble sleight-of-hand that in effect tries to make the public think that the 'trick' was no more than 'regrettable' 'neglect' by the 'IPCC and others' – nothing to do with CRU. In other words, Oxburgh is using a trick to hide the 'trick'. [222]

If the footwork involved in restricting the scope of the inquiry had not been neat enough, when the list of papers was examined, it was realised that far from being those that had been most criticised over the years, almost all were either uncontroversial or even downright obscure. Most of McIntyre's work over the previous seven years had been on CRU's multiproxy temperature reconstructions and not one of these critical papers had been examined. Of the papers that McIntyre had looked at, only Briffa's papers on the divergence problem and Jones' paper on the urban heat island effect – the one that had been the subject of Keenan's fraud allegation – had been examined at all. Yet the latter paper was barely discussed and Keenan's allegations were not even considered worthy of a mention, despite their having appeared on the front page of the *Guardian* just a few weeks earlier. [223]

Amazingly, as the report went on it became even worse: it was revealed that the panel 'was not concerned with the question of whether the conclusions of the published research were correct' but instead had looked at questions of the scientists' integrity and whether there had been any deliberate wrongdoing. In other words the quality of the science in the papers could be abysmal and it would be of no consequence to the inquiry, which was only interested in deliberate dishonesty. This claim in the report flew in the face of repeated statements to the contrary by the principal players from UEA; the university had all along made it quite clear that Oxburgh's review would be precisely what everybody believed it to be – a review of CRU's science. The press release announcing the panel had been unequivocal on the subject:

> An independent external reappraisal of the science in the Climatic Research Unit's ... key publications has been announced by the University of East Anglia. [146]

Acton had strengthened this impression when he told the Science and Technology Committee that allegations of malpractice were the field of Muir Russell and his team, while Oxburgh's panel would look at the science:

> Muir Russell's independent review is not looking at the science, it is looking at allegations about malpractice. As for the science itself...I am hoping, later this week, to announce the chair of a panel to reassess the science and make sure there is nothing wrong.[224]

The impression had been reinforced by Russell, both in his evidence to the Science and Technology Committee and on his panel's website, although we have already seen that there was a rather strange change to the wording at some point.* It seems clear that the university had told Parliament and the public one thing, while Oxburgh had done something quite different.

With the scope of the inquiry so tightly restricted, the rest of the report was almost an irrelevance. There were brief discussions of CRU's work on the land temperature records and paleoclimate, with some mild criticisms made of the unit's failure to devote sufficient attention 'to archiving data and algorithms and recording exactly what they did' and a suggestion that some of the statistical methodologies could have been better. But anyone looking for discussion of the allegations of cherry-icking of data or the use of ad-hoc adjustments would have been disappointed. The determination to give the unit a clean bill of health appeared quite clear – Oxburgh said in an interview that the work of the scientists at CRU was 'squeaky clean'.[225]

THE ADDENDUM

Just hours after the press conference to launch the report, Michael Mann emailed David Hand to take issue with some remarks he had made to the assembled journalists. Hand had been expanding on a passing reference in the report to the use of statistical methods in paleoclimate as practised outside CRU:

> ...inappropriate statistical tools with the potential for producing misleading results have been used by some other groups, presumably by accident rather than design...

---

*See p. 101.

This had been taken by the journalists in attendance as being a reference to the Hockey Stick, and indeed the *Daily Telegraph*'s lead story had been headlined 'Hockey Stick graph was exaggerated'.[226] Hand then had to put up with a stream of emails from Mann complaining about his remarks, and was eventually prevailed upon to issue an addendum to the report. This agreement was sufficient for both Hand and Mann to save face, although Mann then somewhat blotted his copybook by telling a BBC journalist that Hand had 'got his criticism of the stats all wrong' and that he would be issuing an apology. In the event the addendum was merely a reiteration of the text of the report, confirming that the panel was not implying any intent to deceive, and the apology that Mann had said was coming proved to be ephemeral.

THE INVESTIGATION

With the report so brief, it was only a matter of minutes after its publication before the magnitude of the whitewash and the way in which it had been achieved were clear, and sceptics started to turn their attention to fact-checking what was being said. The origin of the list of papers was the first focus and the contradiction between Oxburgh's report – which said that the list had been compiled 'on the advice of the Royal Society' – and UEA's announcement of the panel, which had suggested that Oxburgh's inquiry would examine a list of papers that UEA had submitted to the Science and Technology Committee inquiry. Given that the list that the Oxburgh inquiry had actually examined was almost identical to the UEA list,* Oxburgh's statement appeared on the face of it to be a deception.

After years of trying to extract information from CRU, the sceptic community had considerable expertise in methods for squeezing facts from reluctant bureaucrats, and a series of letters, emails and FOI requests was dispatched over the next few weeks, the latter carefully worded to avoid the battery of tricks that civil servants regularly adopt to thwart the provisions of FOI legislation.

The first responses came from the Royal Society, but these were extremely discouraging. Asked to clarify the role of the society in choosing the panel and the papers, a brief statement was issued by the press office, which failed even to touch upon the question of who had selected the papers for the review:

---

*There was a single difference between the two lists.

> We should all be grateful to Lord Oxburgh and his expert colleagues for a thorough report offering an authoritative assessment of the CRU's research and making clear recommendations. Climate science currently attracts enormous public interest. It is therefore crucial that research sustains the highest standards of rigour and openness. [227]

Well used to this kind of behaviour, the request was reiterated and, after some obfuscation, the Royal Society issued a further statement, which got somewhat closer to a meaningful response but still left much to be desired:

> The Royal Society agreed to suggest to UEA possible members for the Scientific Assessment panel that would investigate the integrity of the research of UEA's Climatic Research Unit...
>
> Members of the panel were suggested on the basis of the excellence of their work and their breadth of expertise and experience (including statistical capability). The Royal Society recommended that the panel had access to any and all papers that it requested and suggested that the review begin by looking at key publications, which were chosen to cover a broad range of subjects over a wide timescale. [228]

This second failure to address the question asked looked suspicious but a further attempt to extract an answer proved fruitless, with the spokesman refusing to say anything more. Another approach was required: a letter direct to Rees, and in due course this elicited a response, although one that was just as evasive as the press office's email. That said, it appeared that Rees was confirming that the society had had no involvement prior to his approval of the shortlist of 13, and no involvement in the panel selection thereafter:

> The panel of members was chosen by the chairman, Lord Oxburgh, from a list of around a dozen approved by the Royal Society. [229]

Rees was also keen to emphasise, however, that he had 'consulted widely' on the subject. He also noted that the list of papers had been chosen by UEA itself, and that he had 'consulted experts' on this topic too. Strangely, these experts had apparently only opined that 'the suggested papers covered a broad range of subjects over a wide timescale'. Again, reference to Rees's correspondence with Davies and Hoskins gives a very different picture of what happened – as we have seen, Hoskins was the only person consulted and in his own words he was not an expert in CRU's science. Moreover, there is nothing in the emails to suggest that Hoskins had said the words that Rees ascribed to him.

Rees' reply raised almost as many questions as it answered and so a second letter was sent, asking about several points, but in particular the identities of the experts consulted on the selection of the papers. Cornered, Rees appears to have decided to opt for discretion, and failed to answer the specific point. He did, however, say that the papers selected were 'representative rather than complete', although, as we have seen, they were not even that. What is more, Rees must have been unaware that a few days beforehand, an FOI request had forced UEA to release Davies' email correspondence with the Royal Society and the truth was out. Rees' dissembling was to come to nothing.

The revelations about the Royal Society's involvement in the deceptions were, however, only the start of it. McIntyre had been busy writing to some of the panellists in an attempt to clarify some of the issues, not least of which was the apparent lack of any formal terms of reference for the inquiry – the report itself had a brief description of what the panel had been tasked with achieving, but beyond that there was nothing. At the start of June, he wrote short letters to Oxburgh and Emanuel asking for the terms of reference and also for copies of any minutes or transcriptions of the interviews, pointing out that the Science and Technology Committee had called for transparency around the details of the inquiries.

The reply from Oxburgh was swift but disappointing. Apparently there were no terms of reference, the arrangements having all been made informally in order to speed up the inquiry, and no records had been kept of the interviews:

> I think that we all felt that looking people in the eye over many hours of discussion about their work and their methods, are just as important as what they say. I believe that the presence of third parties or recording devices could not begin to capture that. It was my judgement that we were most likely to be able to make a fair assessment if proceedings were as informal as possible.*[230]

---

*Oxburgh's suggestion that there had been 'many hours of discussion' is surprising in the light of subseqent revelation of the timetable for the panel's visit – see p. 236.

There was a hint, however, that the instruction to look only for dishonesty, and then only in the published papers, had been imposed by the university:

> The University approached me to chair this review. . . to try to determine whether their staff had been deliberately dishonest in their research activities. [230]

This impression was confirmed by Emanuel, who also seemed to suggest that tight restrictions had been placed upon the panel's scope, ensuring that it did not visit the controversial areas around the IPCC reports, except for the treatment of the divergence problem.

> We were aware that separate inquiries would deal with the IPCC report and with the hacked emails, so we did not get into either of those subjects, though there was some discussion of [hide the decline].

> Our mission was fairly narrowly defined. . . were [there] any improprieties committed in the construction of the papers we were asked to review.*

Another of McIntyre's inquiries led to an extraordinary revelation. It appeared that while being interviewed by the panel, Jones had admitted that it was probably impossible to do paleoclimate temperature reconstructions 'with any accuracy', an admission that flew in the face of repeated assertions by the IPCC authors – including Briffa – that it was 'likely' that modern temperatures were unprecedented. This remarkable news prompted McIntyre to write another letter to Oxburgh, asking that an addendum be made to the report disclosing this shift in position. The response, however, was blunt:

> Thank you for your message. What you report may or may not be the case. But as I have pointed out to you previously the science was not the subject of our study.

> Yours sincerely. . . [231]

Another rich source of information about the background to the inquiry was an FOI request for David Hand's correspondence. Not only did this expose Kelly's notes on the Briffa papers, which, as we have seen, had

---

*Ellipsis in original.

been forwarded to Hand for comment, but also revealed some interesting facts about who knew what.

For example, in one of his emails Hand indicated that he had read *The Hockey Stick Illusion*, by way of briefing himself on sceptics' main complaints. This being the case, he must have been aware of Briffa's involvement in hiding the decline in the Fourth Assessment Report, which is covered in some detail in that book. It is remarkable then that the report referred to 'the IPCC and others' having been less than open about the divergence problem when at least one panel member was aware that this was misleading.

In another email, it was revealed that, extraordinarily, Oxburgh had asked *Phil Jones* for his views on whether the list of papers selected was representative, although it is not clear whether this question was delivered before or after the inquiry itself.

In an intriguing hint that climate scientists and those involved in the inquiries knew more than they were willing to say, an email from Oxburgh to Hoskins on the day of the completion of the inquiry discussed the trick to hide the decline:

> I don't have the IPCC stuff to hand but as a good example, take the WMO brochure ... the cover has a temperature scale between present and 1000AD. Three curves [are] shown – one attributed to Jones, one to Briffa and one to Mann – [there are] no error bands in the illustration and a rather cursory reference in the text inside! We all understood how and why this happened, it's just not fair to blame this on CRU!

Oxburgh has never explained what he meant by these words and there was no discussion in the report of who was actually to blame for the deception of the trick to hide the decline. However, an interview with Phil Jones, recorded in the spring of 2010, gives some further clues. With the inquiries into Climategate already under way, Jones agreed to appear in a TV programme for the BBC's *Horizon* series, in which the UEA alumnus and Nobel laureate Sir Paul Nurse would examine what he called an 'assault' by 'anti-scientific elements'. While it was ostensibly about science as a whole, in fact most of the programme was about climate change, and included a lengthy slot in which Jones was invited to defend himself against the accusations that were still swirling around him.

> NURSE: When Dr Jones was asked by the World Meteorological Organisation to prepare a graph of how temperatures had changed

over the last 1000 years, he had to decide how to deal with this divergence between the datasets. He decided to use the direct measurements of temperature change from thermometers and instruments rather than indirect data from the tree rings, to cover the period from 1960. It was this data splicing, and his e-mail referring to it as a 'trick' that formed the crux of Climategate.

JONES: The organisation wanted a relatively simple diagram for their particular audience. What we started off doing was the three series, with the instrumental temperatures on the end, clearly differentiated from the tree ring series. But they thought that was too complicated to explain to their audience. So what we did was just to add them on, and bring them up to the present. And as I say, this was a World Meteorological Organisation statement. It had hardly any coverage in the media at the time, and had virtually no coverage for the next ten years, until the release of the e-mails.[232]

Although it is impossible to be certain, it appears that Oxburgh felt that Jones was merely acting under orders and should not bear any responsibility for the trick to hide the decline. Of course, since the trick to hide the decline was not part of Oxburgh's very limited remit, there was no reason for him to discuss it in his report. It is nevertheless extraordinary that everybody involved knew how and why the decline was hidden, except the public and their elected representatives.

But perhaps the email that best summed up the scientific inquiry was the one the government chief scientist Sir John Beddington sent Oxburgh – the man he had selected to run the inquiry – shortly after the report had been published. After all the worries of the last six months, Oxburgh appeared to have delivered an outcome that completely met the requirements of the government chief scientist and the sense of a favour having been done is hard to avoid:

Dear Ron

Much appreciated the hard work put into the review, general view is a blinder played. As we discussed at [the House of Lords], clearly the drinks are on me!

Best wishes

John

CHAPTER 9

BOULTON'S REVIEW

With Russell having decided not to take part in any of the interviews of the scientists after the announcement of his inquiry team, when Jim Norton and his fellow panel member Peter Clarke travelled to UEA at the start of March to interview some of the key figures in the inquiry, they went alone. Russell had tasked the two men with interviewing Jones, Osborn and another researcher – Ian 'Harry' Harris – principally on the subject of the CRUTEM temperature set, although they would also cover some of the more serious allegations.[233]

Harris was not a member of the academic staff at CRU but instead was a relatively junior researcher. However, his name was quite familiar to sceptics because of one of the files that had been found among the Climategate disclosures. The 'Harry Readme' file consisted of a long series of notes that Harris had made while studying some computer code bequeathed to him by one of his CRU predecessors. The program related to a surface temperature record called CRUTS, which had been developed at CRU several years previously. In 2006, and much to the consternation of the staff at the unit, a government research body had suddenly decided that they wanted to run the CRUTS code elsewhere and had therefore asked Jones and his team to bring it up to date. Unfortunately the original developer of the code had left the unit and Harris had been given the unenviable job of trying to discover how it worked. The task had proved almost impossible, and in the notes he made as he went along, he had vented his frustration and anger with the poor quality of the computer code and the underlying database. These notes were to be a gift to sceptics, for reasons that are easy to discern from some excerpts:

> ... getting seriously fed up with the state of the Australian data. so many new stations have been introduced, so many false references... so many changes that aren't documented. Every time a cloud forms I'm presented with a bewildering selection of similar-sounding sites, some with references, some with WMO codes, and some with both.

How handy – naming two different files with exactly the same name and relying on their location to differentiate! Aaarrgghh!!

...we don't have the coefficients files...But what are all those monthly files? DON'T KNOW, UNDOCUMENTED. Wherever I look, there are data files, no info about what they are other than their names. And that's useless...

OH F*** THIS. It's Sunday evening, I've worked all weekend, and just when I thought it was done I'm hitting yet another problem that's based on the hopeless state of our databases.

You can't imagine what this has cost me – to actually allow the operator to assign false WMO [station] codes!! But what else is there in such situations? Especially when dealing with a 'master' database of dubious provenance (which, er, they all are and always will be)...This still meant an awful lot of encounters with naughty master stations, when really I suspect nobody else gives a hoot about [them]. So with a somewhat cynical shrug, I added the nuclear option – to match every WMO [station] possible, and turn the rest into new stations....In other words, what CRU usually do. It will allow bad databases to pass unnoticed, and good databases to become bad, but I really don't think people care enough to fix 'em, and it's the main reason the project is nearly a year late.

Plots 24 yearly maps of ...reconstructions of growing season temperatures. Uses 'corrected' [tree-ring data] – but shouldn't usually plot past 1960 because these will be artificially adjusted to look closer to the real temperatures.

It's botch after botch after botch.

The Harry Readme file had not been at the centre of sceptics' attention – being related to computer codes, many people who would otherwise have been interested steered away to the easier issues in the emails. However, many professional programmers had been through the file and had been appalled by what they had seen of the quality of CRU's coding and data management. This was potentially a very serious issue that cast doubt over the whole of CRU's scientific output. However, Harris's explanation of the background to the creation of the Readme file and his assurance that this work had nothing to do with CRUTEM appears to have been sufficient for Norton and Clarke to lose interest in the subject; even an admission that he had been unable to replicate the CRUTS series from the original code and had been forced to recreate the code from scratch apparently left the two panel members unmoved.

With the Harry Readme issue brushed aside, Norton and Clarke decided that they would next question Jones on the CRUTEM dataset. The CRU man explained a little of its history and then went into some detail on the adjustments that were applied to the raw data. He also spoke about the computer program used in the process, noting that this was relatively simple, amounting to only about 100 lines of code. This observation might have been expected to raise the question of whether this short program had any issues analogous to the ones described in the Harry Readme file, but there is no sign that Norton and Clarke probed any further.*

The conversation then moved on to the thorny issue of the Chinese temperature stations and Keenan's fraud allegation.† The questioning again appears to have been very cursory, with Jones stating that he 'did not believe...that there had been gross moves of location of measuring stations, but recognised that there could well have been local moves'. How Jones could hold any scientific view about stations with no recorded histories is unclear, but his defence appears to have relied upon a second paper on UHI, which he had published in 2008, in the aftermath of Keenan's original allegations against Wang. This, like the earlier paper, concluded that the effect of UHI was small. Unfortunately, Norton and Clarke viewed his explanation as adequate and did not question him any further. Had they done so they might have uncovered the surprising truth – that the new paper relied upon the same missing station histories as the 1990 paper that was at the centre of the furore, and was therefore just as unreliable.

The last lines of the minutes record very briefly a moment in which one of the major allegations was at last discussed. The complete minute was as follows:

> Suggestions that e-mails should be deleted
>
> 17. Prof Jones, in response to questioning, noted that he had not received any specific training on DPA/FOIA/EIR issues from the UEA.[233]

It is important to note that as the FOI legislation came into force in 2005 the university had sent Jones a leaflet about the Act,[68] and his

---

*The Harry Readme was actually discussed later on in the minutes – I have reversed the order here for ease of narrative. However, Norton and Clarke would presumably have been aware of the concerns from their preparations.

†See p. 49.

email correspondence made it clear that he was quite aware that the legislation was in force. In fact, in the weeks after the Act came into force, Clare Goodess had sent an email to all members of CRU emphasising the importance of complying with the new law and stating that, when in doubt, FOI requests should be referred to her or to Jones.[234] In addition, a few days after he met Jones and Briffa back in December 2009, Russell had been told by UEA's FOI officer, David Palmer, that Jones and Briffa 'had been made fully aware of their responsibilities under the acts...as part of the process of managing the requests'.*[236] It is clear that if Jones gave the impression that he knew nothing about the provisions of the legislation he was misleading the inquiry. However, once again neither Norton or Clarke appear to have followed up on his response.

Even more damning of the conduct of the inquiry, however, is the fact that Norton and Clarke do not even appear to have asked Jones and the other scientists what they had done after they had discussed deleting emails. This failure is incomprehensible given that the request to delete emails was probably one of the most notorious incidents to emerge from Climategate. Rather than take their opportunity, Norton and Clarke appear to have simply listened to Jones discussing his lack of training before closing the meeting for the day.

RUSSELL AND THE ACCOUNTANTS

Towards the end of the month Russell and Eyton visited UEA for another round of meetings, although these were focused on administrators rather than on the scientists who were supposed to be the focus of the inquiry. The minutes show that the interviews addressed only peripheral issues such as management structures and risk management.[237,238] Once again, there was only a passing reference to one of the allegations from the emails – this time a relatively minor incident in which Briffa appeared to have forwarded cash to the personal bank account of a Russian co-author, Stepan Shiyatov. The request from Shiyatov was quite clear on the reason the money had to be delivered in this way:

> Also, it is important for us if you can transfer the ADVANCE money on the personal accounts which we gave you earlier and the sum for one occasion transfer (for example, during one day) will not be

---

*Palmer warned Jones about deleting emails subject to FOI on 8 December 2008. It was made clear to him in that message that this would represent a criminal breach of the law.[235]

more than 10,000 USD. Only in this case we can avoid big taxes and use money for our work as much as possible.[239]

The transaction appeared highly unusual, although it had not attracted a great deal of comment. When Russell and Eyton questioned the UEA accountant, however, he could reveal little about the particular transaction,which dated back to 1996. Even after he had researched the records thoroughly he could only state that such a cash payment would have required the authorisation of the director of finance and that it appeared to have been recorded in the ledgers.

RUSSELL BOWS OUT

A few days after his visit to UEA with Clarke, Norton was again at CRU, this time accompanied by Russell. The two men were to hold interviews with the UEA staff who had been involved in the processing of FOI requests, but in terms of the investigation their visit appears to have been of little consequence, amounting to little more than an opportunity for UEA staff to outline their procedures. The only interesting snippet of information was a denial by FOI officer Palmer that the university had adopted a policy of refusing requests from anyone associated with *Climate Audit*, but while Norton and Clarke might have been expected to challenge this assertion – perhaps by raising the extraordinary series of reasons given for refusing to give McIntyre Jones' temperature data – in fact the two men let the matter pass, apparently uninterested in discovering the truth.

That said, the visit *was* significant, but for an entirely different reason – it appears to have been the last time that Russell visited the university. This was an extraordinary decision by the inquiry chairman. Although he had met Jones and Briffa for exploratory meetings a few weeks earlier, the two scientists were yet to be subjected to any serious questioning and it seems very surprising that Russell felt that his presence at these crucial interviews was not required. It is hard to imagine that he did not realise that he would be severely criticised for his apparent lack of interest in the mechanics of his inquiry.

BOULTON'S INQUIRY BEGINS

Whatever Russell's reason for ducking out of the remainder of the interviews, responsibility for dealing with the allegations against Jones and Briffa was passed to Boulton, who attended UEA on 9 April, accompanied by Peter Clarke.[240] Remarkably, this meeting turned out to be the only

significant questioning of CRU staff about the major Climategate allegations, and once again the procedures seem to have left much to be desired: Jones and Briffa were interviewed together rather than separately,
and there is little evidence of Boulton and Clarke performing anything
other than a superficial review.

The record of the responses put forward by the two CRU scientists
is derived from a document that purports to record the 'salient points'
discussed, suggesting that it represents an edited account of the meeting
rather than a complete one.[240] Although this document was apparently
corrected for accuracy after the meeting, Jones and Briffa subsequently
appear to have had second thoughts about the story they had agreed,
and they submitted a paper to the inquiry setting out their views in more
detail.[241] Further details of CRU's position were set out in two further
documents: the scientists' formal submission to the inquiry and their
response to a newspaper article about the Yamal controversy that Ross
McKitrick had written shortly before Climategate.[242,243]

EXPLAINING AWAY THE DECLINE

CRU's written evidence to Russell had gone into considerable detail on the
hiding of the decline, outlining the different ways the unit had dealt with
the divergence problem.[244] Although Jones and his colleagues listed a
large number of papers in which Briffa's declining data series was used,
there were essentially only three ways in which they had dealt with it:
showing the divergence, as Briffa had done in many of his papers; truncating it, as the IPCC had done; or truncating it and then splicing the
instrumental temperatures in its place, as Jones had done in the WMO report. Strangely, however, the scientists made no mention of the handling
of the divergence problem in the IPCC reports.

Some of the explanations for these manipulations were bizarre. For
example, it was asserted that the technique of truncating the decline
was used in presentations focused on understanding temperatures of the
past, and it was argued that it was acceptable to hide the decline in these
circumstances because it avoided 'presentation of values that are known
to be unrepresentative of the real temperatures'. They then continued:

> Of course, the recent divergence in these data will be less clear
> if post-1960s values are excluded and that represents a potential
> disadvantage of this exclusion if this divergence is important for
> assessing confidence in the earlier reconstructed values.[244]

This appeared to be a clear admission that hiding the decline hid uncertainties in the reconstructions from the reader, but the CRU team went on to say that when this was an issue they included 'appropriate caveats and references to the articles where the limitations are explored in greater detail'.[244] McIntyre has subsequently described this assertion as a 'fabrication':

> In virtually all cases [of truncation], they included no caveats whatever. To date, I have not identified a single publication in which they explicitly state that they have deleted post-1960 data and why. In some cases, the caption says that the Briffa data is from 1402–1960, but in such cases there is no explicit statement that data was deleted and why. In other cases, there is not even a hint that the data has been chopped back to 1960.[245]

The interview of Jones and Briffa appears to have been largely inconsequential, with Boulton and Clarke failing to challenge any of the assertions made by the two scientists. Jones said that he had no recollection of having sent the hide the decline email, and his verbal defence was largely circumstantial, arguing, for example, that the WMO report was technical in nature and had not been intended for policymakers. He claimed that he had hidden the decline at the behest of his WMO masters, who had wanted to simplify the report for the benefit of their readers. When the question of hiding the decline in the IPCC's Third Assessment Report was raised, Briffa noted that he had not been part of the writing team for that report and according to the inquiry minutes said that others had misused his data. This reply was true if by 'writing team' Briffa was meaning the lead authors. However, it did rather tiptoe around the issue that both Briffa and Jones were contributing authors to the chapter concerned and would therefore probably have submitted text and/or diagrams to the review.[246] In fact, when the CRU scientists were given the opportunity to review the minutes, Briffa disputed ever having said that others had misused his data.[241] Briffa also volunteered that he had fully explained the divergence problem when he had taken the reins as a lead author on the Fourth Assessment Report, although of course he did not explain that he had done this only as a result of protests from sceptic reviewers such as McIntyre and McKitrick, or that he had refused to reinstate the declining tree ring series in the relevant diagram.

Briffa's last line of defence was the one he had used again and again over the preceding months, namely that he had discussed the divergence problem in his papers. Of course, this did not excuse truncating the affected data, something that must have been obvious to scientists as experienced as Boulton and Clark, but there was another intriguing aspect to the hide the decline story that puts the evidence of Jones and Briffa in an unflattering light. Unbeknown to Boulton and Clark, and at that time unknown to anyone outside CRU, there had been other instances of hiding the decline in papers published by CRU scientists, even before the Third Assessment Report. The earliest example was from a paper Jones had published in *Reviews of Geophysics* in 1999 (see Figure 9.1), suggesting that it was the CRU's director who had started the rot, rather than Mann or Briffa.[247]

In the text explaining the graph, Jones had said that the agreement between proxies and instrumental temperatures was at its worst in the nineteenth century, something that was not true – it was just that the considerably greater disagreement in other periods had been deleted. There was certainly no caveat about the reliability of the records, and indeed some of the other statements in the text could only be supported by ignoring the deletion of the decline. For example, Jones' position that no late twentieth century warming could be seen in the proxies because they 'end before the early 1980s' would have been exposed as untrue if the Briffa series, which ended in 1994, had been included.

At around the same time, Briffa and Osborn had published a paper in *Science* in which they had been faced with twin divergences:[248] the twentieth century one that we have already seen and another in the fifteenth and sixteenth centuries, where the data suggested extreme cold, in stark contrast to the other temperature reconstructions. Publishing the data series in full would therefore have undermined the credibility of the whole field of paleoclimatology, something that was clearly out of the question with the Third Assessment Report imminent. A decision appears to have been taken to remove the problematic data and by the time the paper was published, both divergences had been removed (see Figure 9.2).

The details of the two papers from 1999 were, of course, not known to Norton and Clarke, so the two panel members can hardly have been expected to ask the CRU scientists about what had happened. However, with the benefit of hindsight, we can see just how misleading Jones and Briffa were being when they suggested that they had discussed the divergence problem in their papers. In these two important publications, the doubts raised by the anomalous behaviour of Briffa's series had only

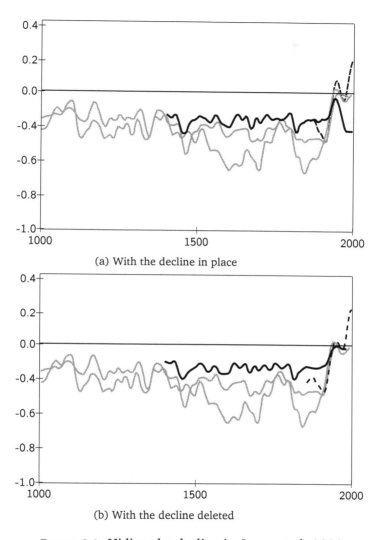

(a) With the decline in place

(b) With the decline deleted

FIGURE 9.1: Hiding the decline in Jones *et al.* 1999

Figures represented are temperature anomalies vs year. Briffa's series is shown in black. The dotted line is the instrumental temperatures.

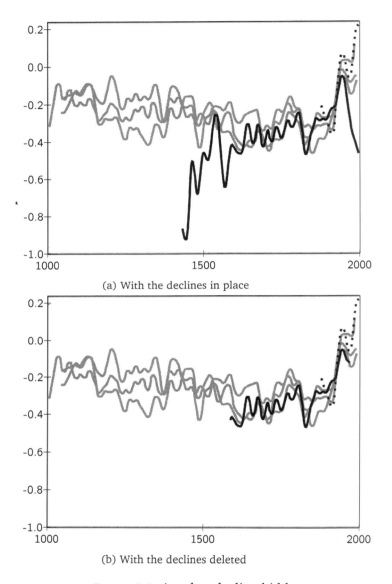

(a) With the declines in place

(b) With the declines deleted

FIGURE 9.2: Another decline hidden

Figures represented are temperature anomalies vs year. Briffa's series is shown in black. The dotted line is the instrumental temperatures.

been alluded to in the vaguest terms and the implications of the divergence problem had gone undiscussed. Instead, Briffa and Osborn had implied that their findings were supportive of the Hockey Stick, the conclusions of which, they said, 'must surely be accepted'.[248] The suggestion that the two CRU men had openly discussed their truncation of the divergence problem was clearly a false one; Norton and Clarke had been tricked.

POLAR URALS AND YAMAL

The release of the Climategate emails had revealed some more intriguing information about the Yamal data, not least of which was the cash payment CRU had made to the Russians' personal bank accounts to fund the work. There was also an email from Tim Osborn to Tom Wigley, which confirmed McIntyre's suspicions that any data he would get from the Russians was unlikely to be the same as that used by Briffa:

> I would really strongly suggest that you contact Keith Briffa about exactly what these series are and what the primary reference to them should be. The reason is that there are multiple versions of... Yamal and the differences are certainly not insignificant![13]

Another remarkable find among the Climategate disclosures was a folder of Keith Briffa's data, containing all the Yamal tree-ring measurements that McIntyre had been requesting for so long. McIntyre was able to deduce that the data had been in CRU's possession since the early 1990s, and this therefore raised some worrying question marks over the conduct of Briffa's colleague, Tim Osborn. In 2006, as part of his efforts to get hold of the Yamal data, McIntyre had approached *Science*, which had published a paper by Osborn and Briffa that used the Yamal series. At the time, Osborn had told McIntyre that he did not have any of the measurement data, explaining that CRU simply picked up the figures provided by the Russians.[249] Now that it was clear that the Yamal data had been there on the CRU servers all along, Osborn's claim looked less than honest but, as one *Climate Audit* commenter noted, if one read his reply carefully, there was perhaps some wriggle room (emphasis added):

> *we* clearly state...
>
> *we* state...
>
> *we* replaced...

we favoured...

we clearly did not use...

we did use...

The source *we* gave for these three series is Briffa (2000).[250]

...and then,

I don't have any core measurement data...

As McIntyre rather acidly observed:

If Briffa had the data in his directory (as opposed to Osborn), isn't this so-to-speak a trick? And not a very good one.[250]

However, perhaps the most important Climategate revelation, at least as far as Yamal was concerned, was an email from Briffa to a scientist at the UK's Met Office, which discussed the regional tree ring chronologies that had been prepared for Briffa's 2008 paper. As we saw in Chapter 2, Briffa had decided not to publish the Urals chronology, instead using the very narrowly-based Yamal series, which had a pronounced hockey stick shape.*

Hi Philip,

We have three 'groups' of trees:

'SCAND' (which includes the Tornetrask and Finland multimillennial chronologies, but also some shorter chronologies from the same region). These trees fall mainly within the 3 boxes centred at:
17.5E, 67.5N
22.5E, 67.5N
27.5E, 67.5N

'URALS' (which includes the Yamal and Polar Urals long chronologies, plus other shorter ones). These fall mainly within these 3 boxes:
52.5E, 67.5N
62.5E, 62.5N (note this is the only one not at 67.5N)
67.5E, 67.5N

'TAIMYR' (which includes the Taimyr long chronology, plus other shorter ones). These fall mainly within these 4 boxes:
87.5E, 67.5N

---

*See p. 56.

102.5E, 67.5N
112.5E, 67.5N
122.5E, 67.5N[251]

The email dated back to 2006, so it had become clear to McIntyre that, just as he had suspected, Briffa *had* prepared a regional chronology incorporating Yamal and the Polar Urals. The rebuttal that Briffa had issued shortly before Climategate, in which he said that he and his co-authors had not considered alternative tree ring series in the area was starting to look thoroughly misleading.[*]

McIntyre's submission to the Russell inquiry had made the importance of the email and the Urals chronology quite clear:

> 'Cherry-picking'
> 13. There has been considerable suspicion that CRU cherry-picked the Yamal chronology over the updated Polar Urals chronology or a still unavailable combined chronology attested in Climategate Letter 1146252894.txt.[252]

However, despite this, it appears that Boulton did not ask Briffa about the Urals chronology; the notes of his interview of Briffa on 9 April 2010 make no mention of it, although there is considerable discussion of the earlier allegations about the paucity of data in the Yamal series.

Some weeks after the interview, Boulton wrote to Briffa to follow up on Yamal, asking him, somewhat strangely, to respond to an article McKitrick had written about the affair the previous year. Because this predated Climategate, McKitrick had made no mention of the existence of the Urals chronology, but had only wondered why such a series had apparently not been created.

Briffa appears to have worded his response very carefully. In one place he repeated the claim that he and his colleagues had made the previous year, namely that they had not even considered the other sites in the Yamal region:

> McKitrick is implying that we considered and deliberately excluded data from our Yamal chronology. The data that he is referring to were never considered at the time because the purpose of the work reported in Briffa (2000) and Briffa et al. (2008) was to reprocess the existing dataset of Hantemirov and Shiyatov (2002).[242]

---

[*]See p. 58.

We should note in passing that the stated purpose of Briffa's 2008 paper was actually to consider regional chronologies in the boreal forest. And of course Briffa had been considering the extended dataset since as early as 2006, so he was clearly misleading Boulton. Then, however, Briffa went on to rather contradict himself by raising the subject of the Urals chronology, but in a way that left the reader rather unclear as to just how much work had been done it:

> ...we had intended to explore an integrated Polar Urals/Yamal larch series but it was felt that this work could not be completed in time and Briffa made the decision to reprocess the Yamal ring-width data to hand [242]

The fact that the new series *had* been created, and did not have a hockey-stick shape, was thus deftly avoided. But it was a secret that would not remain hidden forever.

The rest of Briffa's responses concentrated on the question of the Polar Urals aspect of the cherrypicking allegations. Briffa argued that, contrary to what had been alleged by McIntyre, Yamal was actually a better representation of the region than Polar Urals. [253] Faced with these diametrically opposite conclusions about the same data, Boulton might have been expected to test the competing claims. Assessing the reliability of different tree ring chronologies was a routine task in the field of dendroclimatology, and the dispute between Briffa and McIntyre could easily have been be settled by observing the number of tree ring cores in the two series and testing how well the different cores matched up against each other – a measure known as the replication. As McIntyre had observed, the Polar Urals update was clearly better replicated than Yamal, and should clearly have been preferred. There is, however, no sign that Boulton questioned Briffa's opposite assertion.

Briffa made one major admission about Yamal in his evidence to Boulton, conceding that his failure to discuss Yamal's low core count and high uncertainty 'could be considered an omission'. [240] However, elsewhere he was admitting nothing. He noted that although he had not handed the data over, McIntyre had been able to get it from the Russians who, he said, had published the core count in their paper on Yamal – in other words McIntyre should have known that the core counts were low in the twentieth century. Briffa's defence was interesting, but rather avoided the issue of whether the data he had used was the same as that used by the Russians – there is now no doubt that the two versions were in fact different. It is also noteworthy that the Russians, who had processed the data in a different way to Briffa, had seen no hint in the tree rings that twentieth century temperatures had been unprecedented.[254]

Briffa also contradicted McIntyre's argument that Yamal had been widely used in the temperature reconstructions in the IPCC reports, claiming that it had only appeared in four of the ten reconstructions in the Fourth Assessment Report. This was somewhat misleading since several of the IPCC's reconstructions did not extend back far enough to touch upon the vexed question of the Medieval Warm Period. Of the eight *long* reconstructions, there were actually five that used Yamal – Briffa arrived at a figure of four by disputing that Yamal was used in a paper by Hegerl *et al.* despite the fact that Hegerl had clearly stated that she had.* So when Briffa argued that Yamal was only used in four of ten reconstructions, the truth was that it had been used in five of the eight that were relevant to the question at issue – the existence of the Medieval Warm Period.

These were simple questions and Boulton therefore had ample opportunity to resolve the arguments one way or the other. It is inexplicable that he appears to have made no attempt to do so, and indeed when the report finally appeared it stated that Yamal was only used in four of *twelve* reconstructions in the Fourth Assessment report (a footnote referred to the possibility of a fifth), a proportion that was misleading in the context of the question.

PEER REVIEW

Strange scope, strange parallels

With the issue of peer review having only been touched on in Russell and Norton's earlier exploratory interview of Jones and Briffa,† this second meeting was likely to be the most important opportunity for the inquiry to determine if anything untoward had taken place. The minutes of the interview make it clear, however, that the opportunity was largely spurned. The most important instance of a possible corruption of the peer review process was the sidelining of James Saiers, the editor at GRL, in order to get the Ammann and Wahl comment on McIntyre's paper accepted.‡ The inquiry should have been relatively simple, requiring

---

*The paper states that 'the [west Siberia long] composite involved Yamal and the west Urals composite...'.[255]

†See p.110.

‡See p. 14.

Boulton merely to determine whether any of the CRU scientists had acted upon the suggestion that Saiers might be 'ousted' and, if so, what they had said and to whom. Remarkably, there is no record of these simple questions having been asked. Instead Boulton concentrated on two other sets of allegations.

The first of these concerned the Soon and Baliunas affair. For this part of the meeting, Jones and Briffa were joined by a colleague, Clare Goodess, the CRU researcher who had been part of the editorial team at *Climate Research* at the time of the furore over the Soon paper. Goodess's position at the journal had been mentioned in the Climategate emails* – indeed she had been copied in on some of the messages in which the idea of precipitating a mass resignation of the editors had been mooted.[3] Goodess was therefore in many ways a good choice as an interview subject as she knew almost as much about what had gone on behind the scenes as Jones did, and her involvement would have been obvious to Boulton and his team from a review of the disclosed emails. In fact, her centrality to the plot would have been made entirely plain if Russell had kept to his published work plan and reviewed all of the emails on the CRU server. However, at that point work had not even started on this vital task, and the server and emails remained in the hands of the police officers who were investigating the Climategate disclosures.

The minutes suggest that Goodess did most of the talking in this part of the meeting, reiterating the story that she had already given.[†] Perhaps unsurprisingly, she appears to have made no mention of her involvement in the discussions of journals boycotts and mass resignations. Jones' input to the meeting seems to have been limited to declaring that his reaction had not been 'improper or disproportionate given the self-evident errors of the paper'.[240] Remarkably, the discussion only appears to have touched on the emails sent on the day that the Hockey Team first noticed the publication of the Soon and Baliunas paper: 12 March 2003. The following four months of plotting against von Storch and de Freitas appear to have gone unnoticed, and once again Boulton failed to inquire whether there were any approaches by CRU staff to the journal: the question of whether *Climate Research* had been threatened in some way was simply not discussed.

The second allegation investigated by Boulton under the heading

---

*See p. 3.
†See p. 7.

of peer review did not actually relate to peer review at all. This was the claim that Jones had tried to prevent the sceptic geographer Sonia Boehmer-Christiansen from using her university affiliation. Readers may recall that this allegation had been examined by the House of Commons Science and Technology Committee inquiry, again as if it had something to do with peer review.* Just like that earlier inquiry, the misconstruing of the allegation into one of peer review abuse meant that the outcome could be made a foregone conclusion: Boulton would simply be able to report that he had found no evidence of trying to exert improper influence on Boehmer-Christiansen or her journal. Since that had not been her complaint, this is not entirely surprising.

At this point, the parallels between the Russell inquiry and the Science and Technology Committee inquiry start to look rather interesting. Both investigated the Boehmer-Christiansen allegations as if they were related to peer review and ignored the more recent and arguably more serious Saiers affair. The intrigue is hardly diminished when one recalls that the Science and Technology Committee and the Russell panel also both concentrated much of their effort on the CRUTEM temperature record, which was almost entirely peripheral to Climategate and about which there were no serious allegations. So while there is no evidence that there was any co-ordination between the two separate inquiries, their parallel failings are nevertheless surprising.

The Horton paper

In the wake of Philip Campbell's resignation from the inquiry, Russell had invited Richard Horton, the editor of the *Lancet*, to take up the position of the team's adviser on peer review. Horton had agreed to come on board and his involvement had been reported at the Russell team meeting on 28 April 2010. Horton had been tasked with writing an introduction to the peer review process, covering its history, its efficacy and discussing how it should work in practice. While interesting, this was only peripherally related to the specific allegations and in fact the main body of the final report made little reference to it. However, Russell did note Horton's description of the peer review process as often being heated and hotly contested:

> Authors and reviewers are frequently passionate in their intellectual combat over a piece of research. The tone of their exchanges

---

*See p. 136.

and communications with editors can be attacking, accusatory, aggressive, and even personal. If a research paper is especially controversial and word of it is circulating in a particular scientific community, third-party scientists or critics with an interest in the work may get to hear of it and decide to contact the journal. They might wish to warn or encourage editors. This kind of intervention is entirely normal. It is the task of editors to weigh up the passionate opinions of authors and reviewers, and to reflect on the comments (and motivations) of third parties. To an onlooker, these debates may appear as if improper pressure is being exerted on an editor. In fact, this is the ordinary to and fro of scientific debate going on behind the public screen of science. Occasionally, a line might be crossed.[256]

There was nothing in Horton's description of the way peer review operated in practice that CRU's critics would have found objectionable, and in fact the closing sentence of this excerpt was seen as very important. Horton left no doubt that illegitimate approaches to journals were a reality, if perhaps a rare one. Deciding whether anyone at CRU had crossed the line in this way was presumably therefore going to be a task that Russell and his team would find hard to ignore.

Moreover, if Horton's clarification of the key question for the panel to address had not been important enough, he also raised a question that had previously been overlooked by everyone: there was a possibility that the confidentiality of the peer review process had been breached. As Horton explained:

No manuscript should be passed to a third party by a reviewer without the permission of the editor, usually on the grounds of improving the quality of the critique of the manuscript by involving a colleague in the review process. A disclosure to a third party without the prior permission of the editor would be a serious violation of the peer review process – a breach of confidentiality.[257]

Although Horton did not specifically link his concerns to CRU, there were indeed several instances of what appeared to be breaches of peer review confidentiality referred to in the emails. We have already seen one possible such breach, although since this involved Mann it would not have been in the remit of the Russell panel.* However, there was at least one other email exchange that raised concerns in this area. This message had been sent by Jones to Mann in February 2004:

---

*See p. 13.

Can I ask you something in CONFIDENCE – don't email around, especially not to Keith and Tim here. Have you reviewed any papers recently for Science that say that [two Mann papers] have underestimated variability in the millennial record ...Just a yes or no will do. Tim is reviewing them – I want to make sure he takes my comments on board, but he wants to be squeaky clean with discussing them with others. So forget this email when you reply.

Cheers

Phil[258]

In another incident, Jones suggested to a researcher at the Met Office that he was willing to send them a copy of the manuscript of Wahl and Ammann's paper in *Climatic Change*.

> I have reviewed the [*Climatic Change*] paper by Wahl and Ammann...I can send you the paper if you want, subject to the usual rules.[259]

In another, Mann sent a manuscript to Jones:

> Wahl and Ammann is peer-reviewed and independent of us. I've attached it in case you haven't seen (please don't pass it along to others yet). It should be in press shortly.[26]

Jones later forwarded the manuscript to Osborn and Briffa:

> Just look at the attachment. Don't refer to it or send it on to anybody yet. I guess you could refer to it in the IPCC chapter – you will have to some day.[26]

Another email, sent to Briffa by an Australian researcher, Ed Cook, hints that Mann's email was not an isolated occurrence:

> I got a paper to review...written by a Korean guy and someone from Berkeley...If published as is, this paper could really do some damage. It is also an ugly paper to review because it is rather mathematical...It won't be easy to dismiss out of hand as the math appears to be correct theoretically...Your assistance here is greatly appreciated.[260]

Although the paper does not seem to have been attached to the email, in his reply Briffa makes it clear that he is unconcerned (or perhaps unaware of) issues of peer review confidentiality, saying to Cook, 'from what you say it is a waste of time my looking at it but send a copy anyway...'.

Gatekeeping – Hunzicker and Camill

Cook's email also points at another serious set of allegations related to the area of peer review: the possibility that climate scientists were routinely rejecting papers that questioned the IPCC consensus. Cook's implied wish to 'dismiss out of hand' the paper he was reviewing was one example, but there were others too. In the same email thread, Briffa is seen to say

> I am really sorry but I have to nag about that review – Confidentially I now need a hard and if required extensive case for rejecting – to support Dave Stahle's and really as soon as you can. [261]

The incident was not discussed by Boulton at his interview of Jones and Briffa, and in fact Briffa's defence of his actions – citing his email correspondence relating to the paper in question – only appears to have been sent to Russell a few days before the publication of the report.* [262]

This story related to a paper that Briffa had reviewed in his capacity as the editor of the journal *Holocene*. The paper in question, by Hunzicker and Camill, was a reconstruction of drought history in Montana. One of the reviewers had been very critical of the manuscript, saying that the analyses were good but that the calibration and verification were 'awful'. However, the second reviewer failed to issue his review and Briffa, for one reason or another, failed to chase him up. Thus it was nearly a year later before Briffa sent his infamous email, asking Cook for a 'hard and if required extensive case for rejecting'. In the event, Cook was not so critical, although he said the paper could still be 'justifiably rejected'. Subsequently, Briffa appears to have written to the authors, apologising for the delay and offering them a way out – because of the lengthy delay, he said that he would offer them the chance to rewrite the paper, after which he would publish it without recourse to the peer reviewers.

Between the Climategate emails and the explanation provided by Briffa, Boulton would have been able to verify the story thus far. However, this is not the end of the story. Among the emails on the backup server held by Norfolk Constabulary were a handful that revealed what happened next. On the same day that Briffa had written making his offer to fast-track the paper, he had received a response from Camill, who had been excited about the prospect of finally getting his work into print:

> I have been contacted by the lead author (D. Hunzicker), and he is enthusiastic about resubmitting after substantial revision. [263]

---

*The document is dated 2 July 2010.

Camill went on to note, however, that some of the review comments were based on a misunderstanding about what had been done.

> We feel that we can address several of the reviewers comments about sample size and core inclusion...we both agree that our original statement in the methods unfortunately misled the reviewers into believing that we didn't use cores with missing rings. This is not the case. Second, I wrote the section on sample size, and, unfortunately, I misinterpreted the lead author's description of how he established the chronology vis-a-vis sample size.[263]

For some reason, Briffa failed to mention this email in his letter to Boulton but, more disturbingly, there was *another* email that remained undisclosed. This was dated 26 September 2003, and it was from the editor of *The Holocene* to Phil Camill.

> Dear Dr Camill
>
> Your paper on a 672-year tree-ring [drought] reconstruction is listed as having been rejected by Keith Briffa earlier in the year. He should have informed you of this, although it is possible that there was a lack of communication when he was ill.
>
> Yours sincerely
>
> John A Matthews
> Editor of The Holocene[264]

It is clear then that, with or without the peer reviewers providing Briffa with a hard and fast case, the paper had still been rejected. There is no sign, however, that Boulton uncovered this unhappy ending to the story.

Gatekeeping the IPCC report

As well as the *Holocene* story, McIntyre had listed a whole series of similar cases in his submission to the Parliamentary inquiry, in a document that he later forwarded to Russell. There was plenty for Boulton to investigate:

> Mike...Recently rejected two papers (one for JGR and for GRL) from people saying CRU has it wrong over Siberia. Went to town in both reviews, hopefully successfully. If either appears I will be very surprised...
> Cheers, Phil.[265]

> Mike... I can't see either of these papers being in the next IPCC
> report. Kevin and I will keep them out somehow – even if we have
> to redefine what the peer-review literature is! Cheers, Phil.[266]

Boulton does not appear to have investigated the first message at all, but the second was unavoidable as it was the basis of one of the most important allegations in the emails – McKitrick's allegation that Jones had fabricated a claim of statistical insignificance in the IPCC's Fourth Assessment Report.*

This was clearly a serious matter – indeed it was one of the most serious to emerge from the Climategate emails – but, despite this, the Science and Technology Committee had failed to get Jones to provide the citation and *p*-value that supported his claim that McKitrick's findings were insignificant. Indeed, the committee had failed to examine the issue at all, and there was no mention of the McKitrick or de Laat papers in their report. Undeterred, McKitrick had submitted his version of events to the Russell panel, but amazingly, by the end of Boulton's interview of Jones and Briffa the question had still not been addressed. Instead, Boulton wrote a follow-up letter to Jones asking him to address McKitrick's allegations.[267]

In his reply,[267] Jones admitted writing the email that McIntyre and McKitrick had both cited, saying that he sent it 'on the spur of the moment' and noting that he 'quickly forgot about it'. It had, he said, no bearing on his subsequent behaviour. Jones' defence was in essence that the treatment of the McKitrick and de Laat papers was a collective decision and he said that he had not written the relevant section of the IPCC report, although he added that he agreed with its inclusion. But he went on to make a lengthy and somewhat obscure defence of what had been done, insisting that McKitrick's paper was flawed and making the extraordinary claim that his statement that McKitrick's results were not statistically significant was 'based on the laws of physics' and that it was therefore not necessary to show a *p*-value.

> The pattern of atmospheric circulation-related warming appears similar to the geographical distribution of socioeconomic development. Such similarity makes it impossible to use purely statistical methods to ascribe patterns of warming trends to patterns of socioeconomic development. It remains possible, however, to ascribe patterns of warming trends to atmospheric circulation because its

---

*See p. 134.

186

influence is in accord with the laws of physics and can be detected in day-to-day weather variations, on which timescales socioeconomic trends are infinitesimal. As stated, it is essential to extract the known and understood influences first and then look at the residuals.

There is no need to calculate a p value for a statement that is based on the laws of physics. [267]

Jones' argument was almost entirely devoid of scientific credibility – a statement regarding statistical significance is nothing if not a statement about statistics – and Boulton must have felt he had little choice but to issue a guilty verdict. However, at the start of May, he decided to try a different approach, and contacted the review editor responsible for Jones' chapter in the IPCC report, none other than Sir Brian Hoskins, the Royal Society fellow who had rubber-stamped the list of papers that would be examined by the Oxburgh panel.

Hoskins' advice to the Russell panel was very surprising, as it appeared to completely restate the nature of the IPCC reviews, and in a way that appeared to justify Jones' position. According to Hoskins:

> Irrespective of whether a paper is published in a peer reviewed journal, it is the responsibility of the whole team to assess whether a paper's conclusions are robust and justify its arguments carrying weight in the assessment. These decisions for each draft were taken in plenary sessions of the whole team. [268]

For Hoskins to argue that author teams could decide which papers were valid and which were not was extraordinary, because it flew in the face of the IPCC's own statements on the subject. For example, its overview of the review process stated that:

> All chapters undergo a rigorous writing and open review process to ensure consideration of all relevant scientific information from established journals with robust peer review processes or from other sources which have undergone robust and independent peer review. [269]

Since Jones' statement on the statistical significance of McKitrick's findings had no support in the literature at all, Hoskins was directly contradicting the rules of the IPCC, which had been agreed by its member governments. He was also flying in the face of the IPCC's statement of guiding principles, which showed that dissenting views were meant to be explained in the text of the report:

> In preparing the first draft, and at subsequent stages of revision af-
> ter review, Lead Authors should clearly identify disparate views for
> which there is significant scientific or technical support, together
> with the relevant arguments.[270]

However, this discussion of IPCC procedures was almost completely beside the point. McKitrick's allegation was that Jones' statement on statistical significance was fabricated. IPCC rules and regulations could therefore have no bearing on the truth or otherwise of the allegation – Jones' statement was either supported in the literature or it was not. Since Jones had offered up no evidence to counter McKitrick's claim, the only conclusion that it was possible to draw was that he was indeed guilty of fabrication. But whether Boulton would admit that uncomfortable truth remained to be seen.

TAMPERING WITH THE EVIDENCE

McKitrick's allegation of fabrication was not the only serious matter that Boulton chose to investigate by an exchange of letters rather than a for-mal interview. At the end of February, Holland had sent Russell a lengthy document telling the story of the Hockey Stick, and detailing the intri-cacies of the IPCC review process and the shenanigans over Wahl and Ammann's purported replication of the Hockey Stick.[33] A finding that Briffa had abused his position on the author team to favour the Hockey Team's view of paleoclimate to the detriment of McIntyre and McKitrick would strike a serious blow at the IPCC's credibility.

The publication of Holland's evidence was therefore awaited with some excitement. However, when the first batch of submissions of ev-idence was published on the Russell inquiry's website, Holland's was nowhere to be seen. Shortly afterwards William Hardie, a member of the Russell inquiry's secretariat, sent an email to Holland explaining what had happened:

> Your submission has not yet been published as the Review's legal
> advice is that it could be open to a claim of defamation under
> English law if it publishes the current version of your submission,
> as it makes references to, and comments upon, a large number of
> individuals. The Review, unlike the UK Parliament's Science and
> Technology Committee, does not have Parliamentary privilege.[271]

The suggestion that those submitting evidence could not freely state their allegations was very strange and appeared to undermine the whole basis of the inquiry. Did this mean that the inquiry could only investigate trivial allegations? Holland was mystified – as he noted in his reply he had said little that was not already in the public domain. It was hard to see how Russell could reasonably object. However, he had no means to force Russell's hand and so he reluctantly offered to rewrite those parts of the submission that were worrying the inquiry team.

Strangely, Hardie's response failed to make clear precisely which sections of Holland's submission were of concern, but instead reiterated the concerns about libel and suggested that Holland should provide an email address so that people who were interested could still obtain a copy of his submission. This was all very strange, but Boulton's inquiry was about to become odder still.

Soon afterwards, Boulton began his investigation of the Wahl and Ammann affair, writing to Briffa to ask for his side of the story,[272] and the evidence suggests that considerable efforts had been made to prepare the ground. Holland had alleged that IPCC procedures had been breached with regard to deadlines, openness and the use of registered reviewers. Boulton, however, ruled that much of Holland's complaint was out of the inquiry's scope. As he explained to Briffa:[272]

> The allegation is not whether or not detailed IPCC procedures were followed, that is a matter for the IPCC but whether IPCC procedures were misused to favour one particular view of climate change to the detriment of a credible countervailing view.

Much more serious, however, was the way Boulton treated Holland's evidence. In an annex to the letter to Briffa was a heavily edited version of Holland's submission, with the paragraph numbers removed and the text run together so that the fact of the excisions would not be obvious to the reader. The amount of text removed was remarkable.[273] Holland's original submission ran to 27 pages plus several further pages of appendices, but Boulton had deleted all of the first seven pages and much more besides.

Many of the redactions were inexplicable. For example, an excerpt from one of the IPCC review comments was deleted, as was a quotation from IPCC head Rajendra Pachauri, in which he explained that clear reasons were always given for rejecting review comments. However, some of the other redactions seemed to have a theme to them. Holland had quoted IPCC guidelines that showed that papers for inclusion in the Fourth Assessment Report had to be published or in press by 16 December 2005, but Boulton had removed the comment following it, in which Holland had said that there was 'no doubt' that this was the IPCC's original instruction. This was strange, but in several places Boulton had actually redacted quotations from the emails in their entirety, including more than one on the subject of the deadline. For example, he had removed Holland's quotation of a 2007 email from Jones to Wahl and Ammann, which demonstrated that Jones knew that Wahl and Ammann's paper had missed the IPCC deadline a year earlier:

> Gene/Caspar,
>
> Good to see these two [papers] out. Wahl/Ammann doesn't appear to be in *Climatic Change* online [at] first, but comes up if you search. You likely know that McIntyre will check this one to make sure it hasn't changed since the IPCC close-off date July 2006![274]

Holland had also cited the email in which Wahl made it clear to Schneider that he understood the in-press deadline to be the end of February 2006.*[38] This too had been completely removed by Boulton. However, Boulton's editorial work had led to some complications. In the following paragraph of his submission, Holland had referred back to this email, saying that it 'shows that [Wahl] had understood the clear... instruction that the paper had to be "in press" by 16 December [2005] and was not expecting his paper, written with Caspar Ammann, to be acceptable to the IPCC...'.[33] With that message removed, Briffa was then able simply to note that no email was cited and imply that the allegation was therefore nonsense.[47] The need for Briffa to respond to the evidence that everyone knew the paper had missed the deadline was neatly sidestepped.

When it came to the question of Wahl's illicit input into the IPCC review, Briffa did not deny requesting Wahl's help, but claimed that there were not two sides to the debate, but rather a spectrum of opinion. He said that he had asked for Wahl's help as 'knowledgeable, objective and entirely frank arbiter'.[47] Quite how Briffa thought Wahl could be an objective arbiter of a dispute in which he was one of the protagonists remains unclear, but again his representations, however incredible, appear to have been accepted without question by Boulton.

---

*See p. 24.

The main thrust of Briffa's response was that he had tried to be fair to both sides, and he quoted in his defence two emails that he had sent that suggested that even-handedness had indeed been a concern in his mind. He also said that the inclusion or exclusion of the Hockey Stick in the final report would have made little difference to the final assessment of the climate of the last thousand years. This was somewhat disingenuous, not only because the divergence problem threw all of the paleoclimate reconstructions into doubt, but also because the argument about the Hockey Stick was as a result of its having been promoted far beyond its importance as a single temperature reconstruction. The IPCC could not back down on its promotion of the Hockey Stick without losing credibility – the assessment of the climate of the last millennium had very little to do with it.

Despite his protestations, Briffa did make some important clarifications, not least of which was his admission that the version of the Wahl and Ammann paper that was used by the IPCC was not the final one. By the letter of the IPCC guidelines, this meant that it should not have been used in the report, although Briffa implied that because the differences were minor, it actually did not matter. He also did not dispute that the Wahl and Ammann paper relied upon the GRL comment, which had not been accepted by that journal until long after the IPCC deadline. However, he also said that the IPCC authors could not be expected to know that changes would be made to the Wahl and Ammann paper after its formal acceptance.

Perhaps more important, however, were Briffa's observations about the Hockey Stick itself. Boulton's questions suggested that he had entirely misunderstood what the Wahl and Ammann paper was about, suggesting in his covering letter that it was a rebuttal of McIntyre and McKitrick's 2003 paper in *Energy and Environment* rather than of their 2005 paper in GRL. Briffa seems to have been rather taken aback by this error and queried it with Boulton, who confirmed, incredibly, that he was only interested in the 2003 paper. However, Briffa did at least volunteer the important information that the IPCC had *not* disputed McIntyre and McKitrick's observation that the Hockey Stick had failed its verification statistics. This was a significant statement, since it accepted one of the principal criticisms of the Hockey Stick – a criticism that had long been thought to have been rejected by the IPCC. It is therefore worth examining the wording of the Fourth Assessment Report in detail:

> McIntyre and McKitrick... raised further concerns about the details
> of the Mann *et al.* (1998) method, principally relating to the inde-
> pendent verification of the reconstruction against 19th-century in-
> strumental temperature data and [principal components analysis].
> The latter may have some theoretical foundation, but...

This sentence rather gives the impression that the concerns over the verification statistics were *without* foundation, so Briffa's new explanation of its meaning puts the IPCC's conclusions in a completely different light. It suggests that the original text was deliberately left vague, allowing Mann and the Hockey Team to claim vindication while allowing the author team room to issue plausible denials at a later date. One question remains unanswered, however: if Briffa recognised that the Hockey Stick had failed its verification statistics, why did he want to cite it in the report at all?

Briffa had made a strong case that he had been trying to be fair to all sides, but the substantial allegations – that his approach to Wahl and his use of the unpublished comment had represented a breach of IPCC rules and that the deadline had been rewritten – remained problematic. Briffa could in essence only protest that everything he had done was within the rules. It was therefore necessary for the inquiry to get an outside opinion on exactly what was and was not permissible.

Briffa had already made steps in this direction himself, and his evidence to Boulton included excerpts from a statement that he said he had obtained from the IPCC TSU. Briffa said that it had been 'prepared in consultation with the former Co-Chair and TSU of [Working Group I for the Fourth Assessment Report]' – so it is likely that it came from Solomon and Manning.

The TSU certainly seemed to be trying to defend Briffa's actions, but their version of events also contradicted known facts. They said that the deadline for inclusion of new papers had been set at 24 July 2006, without discussing the previous final deadline at the end of 2005. They went on to discuss the need to have a clear cutoff date, stating that this was important to ensure that no new issues were raised in the final draft of the report. They then noted that during July 2006 new papers had been sent to them for possible inclusion and that these had been sent out to chapter lead authors shortly afterwards. This was true, but avoided the interesting subject of why the Wahl and Ammann paper had been sent out in this way, but not the Wegman or NAS reports that McIntyre had suggested.

Finally, they discussed Briffa's contact with Wahl, suggesting that this was quite permissible under the rules of the IPCC:

> The Procedures for preparation of IPCC Reports state that 'The Co-ordinating Lead Authors and Lead Authors selected by the Working Group/Task Force Bureau may enlist other experts as Contributing Authors to assist with the work.' Thus they explicitly allow for contributions by experts who are not part of the full author team. If such an expert is consulted, it is up to the authors to decide if their contribution is significant enough to warrant their designation as Contributing Authors.[47]

The citation given to the IPCC procedures is correct,[270] appearing within the section on appointment of lead authors. The idea that this is what had actually happened in this instance defies belief, however. The IPCC rules quoted suggest that the author *team* were able to bring in outside help on particular sections, not that an individual author could do so, and it is hard to see Wahl's contribution as being by way of helping with the drafting. Finally, the TSU's suggestion that the authors could decide whether to designate the outside expert as a contributing author or not appears to have been invented entirely to assist Briffa. It appears nowhere in the IPCC guidance.

Briffa also obtained support for his case from Eystein Jansen and Jonathan Overpeck, reproducing comments they had made earlier in the year. Jansen said he could see no IPCC rules that prevented Briffa from contacting Wahl, while Overpeck went further, stating that Briffa 'was encouraged to interact with scientific colleagues to obtain any additional information needed to ensure that disparate views of the scientific community were understood'.[47]

Briffa's solicitation of official support for his position had given Boulton a great deal to work with but, it seems, not enough. At the start of June, Boulton decided to conduct a telephone interview with Professor John Mitchell who, as we have seen, was the review editor on the paleoclimate chapter, and was deeply implicated in the Climategate scandal, having given false responses to Holland's FOI requests and possibly even having deleted emails in order to thwart them.*

Mitchell stated in his evidence to Russell that he only became aware of the concerns over whether Wahl and Ammann had met the deadline 'after the event', a claim that appears somewhat odd in the light of their

---

*See p. 42.

having being raised by the expert reviewers of the Second Order Draft.* It is possible that Mitchell had forgotten about having seen these comments at the time or even that he had missed them during his own review, but there are undoubtedly questions raised over his failure to address the issue at the time.

Mitchell's position was that it was permissible to use unpublished material under the IPCC's rules, and that therefore discussion of deadlines was beside the point. This was again surprising, since it contradicted the IPCC's own statements on the subject, which made it abundantly clear that material to be cited had to be published by the official deadline.† It is perhaps significant that Mitchell did not actually cite any written IPCC rules to support his case.

Mitchell's other claim was that there was nothing in the IPCC rules to prevent Briffa from seeking Wahl's input, although again, had Russell investigated slightly further, he would surely have noted the contradiction with the IPCC's own statements on the subject.

Boulton now had a number of statements from people involved with the IPCC declaring that Briffa's conduct had been within the rules. The rules themselves were telling a rather different story. He would have to decide whether the truth lay in the text of the rules themselves or in interpretations of them by by IPCC insiders.

FORENSIC ANALYSIS?

The Russell panel's terms of reference required it to determine whether there had been 'evidence of manipulation or suppression of data which is at odds with acceptable scientific practice and may therefore call into question any of the research outcomes'. In fact the terms were very specific about what Russell would examine, listing 'the hacked e-mail exchanges, other relevant e-mail exchanges and any other information held at the Climate Research Unit'. In view of this, it is perhaps surprising that it was not until the beginning of March that the panel started to consider the other emails – emails that we now know included such important details as the proof that CRU scientists had been using memory sticks, that Jones had met von Storch to discuss the Soon and Baliunas paper, and that Briffa had in fact rejected the Hunzicker and Camill paper. Unfortunately, it was not a straightforward task to get hold of these emails, as

---

*See p. 30.
†See p. 19.

the server, known as CRUBAK, had been seized by Norfolk Constabulary as part of their investigation of the Climategate disclosures. Russell's computer specialist, Jim Norton, was therefore assigned the task of liaising with the police to get a copy.

Norton's first email to the police sets out what the team were trying to achieve:

> ... we would like to examine how the leaked e-mails relate to the rest of the material on the server, e.g. how they were selected and using what search terms, and whether the server still contains any traces of e-mail or data deletion. [275]

The response from the officer in charge of the investigation, Detective Superintendent Julian Gregory, was disappointing. As Gregory explained, the sensitivity of the data on the server meant that to hand over a full copy to the Russell team was impossible. However, he said that if Norton explained in more detail the questions he would like answered, it should be possible to get Qinetiq, the IT specialists who were handling the forensic work for the police, to provide some answers.

After conferring with the rest of the Russell panel, Norton was able to provide a list of three questions, which he emailed to Gregory at the end of March:

> 1. Could Qinetiq please quote costs and timescale just to extract the e-mails from the backups of four specific machines – those of Prof. Phillip [sic] Jones, Prof. Keith Briffa, Dr. Tim Osborn and Dr. Mike Hulme?
>
> 2. Have Qinetiq been able to determine the search terms that were used (either locally on the server) [redacted] set of e-mails that were disclosed?
>
> 3. Would Qinetiq be prepared to run some searches for us on the server as a whole against search terms that we would provide? [276]

In a followup email, Norton also explained that the panel's interest in the other emails on the server was not purely about investigating the actions of Jones and his colleagues. It appeared that the panel were also worried that there might be further disclosures of emails in the future and that these might lead to criticisms of the panel's report.

> We would like to know both, if there are other pointers to areas we should also investigate in the rest of the CRUBAK material, and

whether there is further information out 'in the wild' that might subsequently be released in an effort to denigrate our report once published.[276]

Despite the Russell inquiry being nominally independent of UEA, in practice the inquiry was dependent on the university in a number of areas, and another month passed while Russell sought authorisation from Acton for the cost of extracting the emails. It was therefore not until 20 April that the go-ahead from UEA was finally obtained. Strangely, the questions about search terms were no longer in evidence, with a more restricted work plan now being requested – essentially Qinetiq were only to be asked to obtain the emails of the three principals at CRU– Jones, Briffa and Osborn.

While the negotiations with the police and Qinetiq had been ongoing, it had also been considered necessary to appoint someone to analyse the data when it finally arrived. The minutes of the Russell panel's meeting at the end of April reported that the appointment had been made, although without naming the person who was to fill the position, who was described, slightly strangely, as a *'trusted*, independent, forensic analyst' (emphasis added). It was only with the publication of the report that the analyst was revealed to be Professor Peter Sommer of the London School of Economics, an expert in cyber security and computer forensics.

The data was finally delivered to Norfolk Constabulary in the middle of May, and it might have been expected that the Russell panel would have been able to start the long task of examining the remaining emails soon thereafterwards. However, in the event things were more complicated. Soon after taking the decision to have the emails extracted, the panel appear to have developed concerns over data protection issues and although they retained their own legal advisers, for some reason they then decided to get an opinion from the *university's* lawyers as to what steps they should take.[277] They were advised, surprisingly, that 'unconstrained access to the contents of e-mails on the server ... would raise potential privacy and data protection issues'.[278]* Unconstrained access might well have raised such issues, but the constraints that the university then imposed on Sommer were so extraordinary that within a matter of days he had all but given up.

---

*The university has subsequently resisted attempts to obtain copies of this legal advice under FOI.

Sommer gained access to the data on Friday 14 May 2010. On the following Monday he issued a preliminary report to Russell, in which he explained the problem:

> The material has been given a very high level of security classification which requires that I work at secure facilities and follow particular protocols which, for example, preclude computers being left to run unattended or overnight and at weekends. These procedures, while providing a very high standard of protection to the data, are also very time consuming, particularly in the light of the need for the Review to conclude its work in a timely manner.[279]

If these difficulties were not bad enough, Sommer was also having difficulties with the sheer volume of the material extracted from the CRUBAK server, which he estimated to be 425 times the amount contained in the Climategate zip archive, as well as his own unfamiliarity with the material. Because of all these problems, he explained, it would take 'at least several weeks' to identify and review any emails that might shed further light on the allegations made against Jones and his colleagues.

> It would be for the Review Team and the University to determine whether the cost, inevitable time delays and (at this time) uncertain outcomes could be justified. Until the material is subjected to a much downgraded security level, the likely position will be that the University and its appointed team will not be able to carry out any meaningful analysis.[279]

Sommer's 'Preliminary Report' was in fact his last involvement in the Russell inquiry. As the panel subsequently reported, a decision was taken that this element of the inquiry would not be pursued further, 'on grounds of both time and cost against likely results', although they later claimed that they only saw the emails as 'pointers' for their investigations.* It appears that because of a delay in making a start, and also because of the restrictions placed upon Sommer's working arrangements by the university, an important part of the panel's work was to be set aside. Another stone went unturned.

---

*It is hard not to notice the parallels with Oxburgh's suggestion that the list of papers examined was only a 'starting point' for their inquiry.

OVERPECK'S SPREADSHEET

David Holland's tireless investigation of the back story of the Fourth Assessment report had not stopped just because of Climategate. In fact he had been thoroughly re-energised by all the revelations and his work had continued apace while the Science and Technology Committee had been at work. One of the issues that had particularly attracted his interest was the spreadsheet that Overpeck had sent to the author teams in the middle of July 2006 – this contained a list of the new scientific papers that he had deemed acceptable for citation in the report, including the Wahl and Ammann paper but not the sceptic-friendly NAS and Wegman reports.* However, the Climategate emails contained only Overpeck's covering email and not the spreadsheet attachment, so Holland could only guess at these details. The only way to prove the point one way or the other was to ask for the attachment from CRU via another FOI request.

Perhaps inevitably this new approach had been refused, with the university offering up the twin grounds that the informationwas 'not held' and also that its disclosure would harm the police investigation. As the university explained in their refusal letter, the only place the emails existed was the backup server that had been at the centre of the Climategate story and was now in the possession of Norfolk Constabulary:

> ... the only location that this information was held was on a backup server as the original information had been 'deleted' some years ago. [280]

On the face of it, this appeared to be a valid response – presumably the university could not release information if it was only on the server and there was also no doubt that the emails were subject to a police investigation. It also seemed to confirm everyone's assumption that the CRU scientists had deleted all the incriminating emails. Because of this, and because he was busy with other matters, Holland decided not to appeal the university's response. His decision not to do so, however, may subsequently turn out to be very significant.

Although the Russell panel's investigation of the backup server was ultimately something of a farce, their decision to follow this line of inquiry would have immediately presented Acton and Davies with a problem: among the correspondence on the backup server was the Palutikof email.† This represented clear and unequivocal evidence that CRU sci-

---

*See p. 32.
†See p. 42.

entists had concealed emails – a criminal offence. Although the immediate cause of the decision to use memory sticks – Holland's request for
Briffa's IPCC correspondence – was time-barred, the new FOI request –
for the Overpeck spreadsheet – was not, having only been refused a few
weeks earlier at the end of January. If the Palutikoff email became public knowledge, it could well lead to a further ICO investigation of CRU
and this would almost certainly expose the concealment of the Overpeck
spreadsheet. Such a revelation could only have one outcome: a criminal
prosecution of UEA staff.

The situation was perilous and decisive steps were clearly needed to
save the situation. On 26 March, just ten days after the Russell panel
took the decision to proceed with the investigation of the backup server,
Holland received a surprising new email from UEA:

> Further to my letter of 26 January 2010 responding to your request
> of 28 November 2009 for a spreadsheet sent as an attachment to
> an email of 28 July 2006 from Jonathan Overpeck, I am writing to
> advise you of recent developments in this matter.
>
> By my letter of 26 January, you were informed in good faith that
> the requested information was not held at the time of the request...
>
> It has come to the University's attention that the University does
> hold the spreadsheet in question otherwise than on the backup
> server. We only made this discovery very recently.
>
> Accordingly, I enclose with this letter a copy of the spreadsheet in
> fulfilment of your request of 26 January 2010.[281]

The letter hints at the truth: that the spreadsheet had been concealed
along with the emails on a memory stick, but in fact this clue passed Holland entirely by. Since receiving the original letter of rejection from UEA
he had written directly to several IPCC officials, including Overpeck, to
ask them for a copy. So when UEA suddenly sent it to him, he simply
assumed that Overpeck had resent it to Briffa. It wasn't until a few days
later that Holland's suspicions were raised when the Information Commissioner's office told him that the spreadsheet had been found on a
memory stick.[282]

Holland was not the only one whose attention was seized by the sudden discovery of the emails. The ICO staff who were investigating Holland's complaints against UEA also wanted an explanation, and shortly
afterwards they sent an enquiry to UEA.

The university's response is a remarkable document. According to David Palmer, the UEA FOI officer, when his request had been refused, Holland had requested an internal review of the decision – the first step in the appeal process.

> On receiving notification of Mr Holland's request for an internal review, the Director of Faculty Administration, Michael McGarvie, searched a memory stick, which the University had come to understand might contain additional information, to see if the information could be located by those means. [283]

According to Palmer, the search had revealed Overpeck's covering email and a subsequent search of Briffa's data had unearthed the spreadsheet itself.

There was, however, a significant problem with this story: Holland had not in fact requested an internal review at all. As we saw, he had thought the chances of success were so slim as to make the effort pointless. In fact there were no grounds at all for the university to reopen the request in this way – their own disclosure records noted that Holland's request had been refused. However, for the time being the university's prompt action appears to have been enough to save them – the Information Commissioner made no further efforts to investigate and the trail became quiet. Whether the university had got away with it or whether they had merely bought themselves some time remains unclear: sending false information to a statutory investigation of an allegation of criminal offences is a course of action that has potentially serious consequences.

LATE PREPARATIONS

Russell had originally scheduled the publication of his report for the Spring of 2010, so by the summer there was a strong sense of expectation. Records of the panel's meetings continued to appear on their website, and these were eagerly examined by sceptics keen to see what was going on. The minutes of the April meeting in particular raised some considerable concerns.

> *Review of* IPCC *processes and procedures*
> The Review Team emphasised that it had not been commissioned to examine the IPCC's processes. As part of its remit, the Review is reviewing the CRU's policies and practices for acquiring, assembling, subjecting to peer review and disseminating data and re-

search findings, and their compliance or otherwise with best scientific practice. The Review recognised that a separate independent review of the IPCC's processes and procedures is being conducted by the InterAcademy Council. [284]

This statement could have been read in one of two ways. The innocent interpretation was that the Russell panel did not view questioning the IPCC's procedures as part of their remit. However, the wording was vague, and an alternative, and rather more sinister intepretation was that Russell was signalling that any allegations relating to the IPCC would be ignored. McIntyre quickly sent off a letter to Russell asking him to consider the need to address the question of misconduct by CRU employees as part of their work on the IPCC reports, and to amend his terms of reference – as he was able to do – in order to bring this within his remit.

As McIntyre explained to his blog readers, the inquiries to date had spurned the opportunity to investigate these matters, so it was necessary to make sure:

> The Oxburgh panel already failed their obligation to consider IPCC activities by CRU employees, presuming that this would be done by Muir Russell.

> If the Muir Russell panel fails this obligation as well, the losers will, of course, include those members of the public who had hoped that the 'inquiries' would inquire. But the greater losers would be the climate science 'community' itself, which will, for a day or two, proudly display the worthless baubles of another 'inquiry' that didn't do its job. But such preening with worthless baubles will hardly re-assure the public who expect more from people who purport to be scientific leaders. [285]

It wasn't only McIntyre who was trying to make points late in the day. Just before the end of June the Hockey Team sent in its own supplementary submission, in the form of an email signed by all the main players.

The message began by discussing sceptics' criticisms of the membership of the Russell panel and went on, once again, to present themselves as victims of a plot:

> As climate scientists, we are, regrettably, all too familiar with these tactics. The unfortunate reality is that, to research climate issues

today – at least if one's research findings tend to support human-caused climate change – means to live and work in an environment of constant accusations of fraud, calls for investigations (or for criminal prosecutions), demands for access to every draft, every intermediate calculation, and every email exchanged with colleagues, daily hate mail and threats, and attempts to pressure the institutions that employ us and fund our research.[286]

The letter continued in a similar vein, declaring that 'Not all the evidence submitted to the [inquiry] comes from parties with genuine interest in furthering scientific understanding'. The Team continued by asking Russell to make a number of statements in his report: that scientists' correspondence should be exempt from FOI and that the global warming consensus was unaffected by what had happened at CRU. It was going to be impossible for Russell to please everyone.

THE REPORT

The Russell report finally appeared on 7 July 2010, with a whimper rather than a bang. A few scraps were thrown to the sceptics, but for the most part Russell and his team declared that they had seen nothing untoward. 'We find', said Russell of the CRU staff, 'that their rigour and honesty as scientists are not in doubt'. In addition, the Hockey Team's request that Russell should make a statement that the global warming consensus was unaffected seemed to have been met in full, with one of Russell's three headline findings reading as follows:

> ...we did not find any evidence of behaviour that might undermine the conclusions of the IPCC assessments.[287]

There was little surprise among the sceptics. It had been clear from the moment the Russell panel had been announced that there was little likelihood of a meaningful investigation. However, it was still interesting to see how the mechanics of issuing such an unequivocal exoneration had been managed in the face of such overwhelming evidence to the contrary, and the report was carefully dissected over the following hours and days.

CRUTEM

Beddington's requirement that the inquiry look at the CRUTEM temperature record must have given Russell something of a headache, since he had to come up with some reason for spending so much time and effort on it rather than directing his team towards the actual allegations. The justification that Russell came up with was, to say the least, very thin, declaring that there had been a broad allegation that:

> ... CRU adjusted the data without scientific justification or adequate explanation. Some allegations imply that this was done to fabricate evidence for recent warming.[288]

He then went on to note that he and his team had engaged in a certain amount of construction in order to arrive at this interpretation of what the allegations were:

> While very few of the submissions to this inquiry make the allegations above explicitly, they are nevertheless implied.[289]

This was something of an embarrassment, and subsequent to the Russell report's publication, McIntyre spent some time trying to discover whether there really were submissions of evidence that implied fabrication of recent warming. Searches by *Climate Audit* readers turned up little to support the idea. It appears that Russell felt obliged to make this statement simply to cover up his 'independent' inquiry's blushes at being directed onto peripheral matters by a government adviser. The remainder of this section of the report covered a reworking of the CRU analysis, which, as expected, concluded that the series was soundly based and easy to reproduce, although it sidestepped the awkward question of how the inquiry had managed to do this without a list of the stations used.

Peer review

In his report, Russell characterised the allegations of peer review abuse as follows:

> CRU made improper attempts to influence the peer review system, pressuring journals to reject submitted articles that did not support a particular view of climate change.[290]

Of itself, this characterisation was reasonable, although it did not mention the additional possibility of repercussions for journals that refused to toe the line. However, after this reasonable start, it became clear that the inquiry left a great deal to be desired. Richard Horton's advice that there was an ethical line that could be crossed when scientists issued complaints to journals appears to have been largely ignored.

The Saiers affair was ignored entirely, and the investigation of the Soon and Baliunas affair was cursory at best, with no inquiries made beyond Boulton's brief interviews of Jones and Goodess.

The impression of the investigation of the Soon and Baliunas affair appears even more culpable when one considers the subsequent revelation that Jones had met with von Storch at the end of April 2003 to discuss the paper – Boulton's inquiries do not even appear to have elicited the fact of that meeting having taken place, let alone determined the details of what had been said. There is also no sign that questions as simple as 'Did you contact *Climate Research* about the Soon paper' were ever asked.

In a similar manner, Horton's pointing out that peer review confidentiality was an important issue garnered no reaction from the Russell panel. For them, Jones' word that he had done nothing wrong was sufficient.

Russell's handling of the Hunzicker and Camill paper was equally interesting. Although Briffa had told Boulton that he had not blocked the paper's publication irrespective of merit, no attempt appears to have been made to follow up on this statement, and the report declared that there was something of a mystery over the eventual fate of the paper:

> We can find no evidence that the article was subsequently published but the evidence above demonstrates that the possibility of publication was not rejected.

This inability to determine the eventual fate of the paper is surprising, since Boulton could presumably have simply asked Briffa if the paper had subsequently appeared in *Holocene*. The later revelation that Briffa did in fact reject the paper, but failed to tell Boulton, and the fact that the CRUBAK server contained the emails revealing this fact but that the Russell panel failed to examine it, paint a disturbing picture of the nature of the inquiry.

Fabrication

It had clearly been problematic for Russell and his team to address the allegation that Jones had fabricated a claim in the IPCC report – namely that McKitrick's findings on temperature trends were statistically insignificant – and their clearing of the CRU man relied on a great deal of obfuscation and sleight of hand. Firstly Russell and his team team decided to rewrite the allegation:

[the conclusion on fabrication] depends upon the implication that the response to [McKitrick and Michaels] in the published Chapter 3 was not scientifically credible. [291]

This of course sounded plausible, but was of course not true. The assessment of whether Jones' claim had been fabricated depended on whether it could be found in the scientific literature; the *scientific credibility* of Jones' claim was besides the point and was for a scientific journal to determine through the normal peer review process.

Having established favourable ground in this way, the Russell panel simply pointed to the evidence provided by Hoskins to argue essentially that IPCC authors were the arbiters of scientific credibility and that they could include or exclude anything they liked. The fact that Jones had fabricated a statement in order to justify such a decision was apparently considered irrelevant – the Russell panel was arguing that IPCC authors could say anything they liked in the report, with or without support in the scientific literature.

Wahl and Ammann and the subversion of the IPCC process

It was a similar story on the Wahl and Ammann affair, with the actual allegations being largely sidestepped. Instead of looking at what the rules were and whether they had been broken, the Russell report instead decided to address the question of whether Briffa's conduct had been *reasonable*. Even then it required a certain lightness of touch to avoid some of the difficult questions. For example, on the tricky question of whether it was reasonable for Briffa to approach Wahl, the report noted Briffa's claim that Wahl was 'a knowledgeable and objective arbiter' but neatly avoided questioning whether this was in fact a credible claim. Instead, much was made of the evidence of Mitchell and the TSU and their protestations that everything Briffa had done was within the rules.

So, when it came to whether the inclusion of Wahl and Ammann's paper was permitted under IPCC rules, the question was avoided in the answer, with Russell declaring that it was a

> reasonable attempt to use up to date information that might resolve an issue. [292]

Email requests and FOI

The correspondence between Briffa and Wahl was closely linked to the related question of whether anyone at CRU had in fact deleted emails. At the time of the Russell report's publication, the information that CRU staff had moved all their sensitive emails onto memory sticks was not widely known – outside CRU and the senior echelons of UEA it is possible that only Russell and Norton were aware of the facts, although Holland had his suspicions too.*

There is, however, no doubt that Russell knew, because he had been told that this is what had happened by Jonathan Colam-French. It is therefore inexplicable that the Russell report was virtually silent on the question of deleting emails: there was no mention of memory sticks or 'safekeeping' or anything of the sort. In fact the only statements that touched on the issue at all were a brief note that there appeared to be clear evidence of incitement to delete, followed by the extraordinary statement that the team had 'seen no evidence of any attempt to delete information in respect of a [FOI] request already made'.[293] McIntyre described this claim as 'blatant misrepresentation'. Given the fact that Jones' email to Mann asking him to delete his IPCC-related correspondence was sent just hours after Holland's FOI request and moreover carried the title 'IPCC & FOI', this was a not-unreasonable assessment of Russell's claim. In the light of the subsequent public disclosure of the use of memory sticks and in the knowledge that Russell was previously aware that this had happened, the misrepresentation appears to be highly culpable.

Elsewhere in the report, the findings were slightly stronger, with Russell stating that there was 'evidence that emails might have been deleted in order to make them unavailable should a subsequent request be made for them [under Freedom of Information law]', but no detail was given. He was also critical of the culture at the university, saying that they were unhelpful in response to requests, but it was clear that there was going to be little of substance on the subject of FOI.

Of course UEA were also aware of the use of memory sticks, since it had been Jonathan Colam-French, the university's IT director who had told Russell about what had happened. Despite this, when the university issued its official response to the Russell report a few weeks later, it played along with the misleading finding about email deletions.

> ...we welcome the finding that there was no attempt to delete information with respect to a request already made. This confirms assurances already given to the Vice-Chancellor by colleagues in

---

*See p. 198.

CRU that they had not deleted material which was the subject of a request. We have underlined that such action would have been one of the key elements necessary to constitute an offence under Section 77 of the FOIA and Section 19 of the EIR, the others being that information had actually been deleted, that it was deleted with the intention to avoid disclosure and that it was disclosable and not exempt information. Professor Jones has commented that, while emails are cleared out from time to time, this is to keep accounts manageable and within the allocated storage.[294]

There can be little doubt that the university and at least some members of the Russell panel team all knew what was going on from the very beginning of the Climategate affair. Yet in their official pronouncements they repeatedly gave the impression that nothing untoward had happened. It was another remarkable performance by all concerned.

The Trick

Elsewhere in the report it was more of the same, with allegations of cherrypicking and bodging brushed aside on the vaguest of pretexts. Only on one of the serious allegations did Russell and his team find themselves unable to issue a full exoneration. On the subject of the trick to hide the decline they issued what was to be virtually their only 'guilty' verdict.

> On the allegation that the references in a specific e-mail to a 'trick' and to 'hide the decline' in respect of a 1999 WMO report figure show evidence of intent to paint a misleading picture, we find that, given its subsequent iconic significance (not least the use of a similar figure in the IPCC Third Assessment Report), the figure supplied for the WMO Report was misleading. We do not find that it is misleading to curtail reconstructions at some point per se, or to splice data, but we believe that both of these procedures should have been made plain – ideally in the figure but certainly clearly described in either the caption or the text.[295]

Russell's statement was a striking contrast to the Science and Technology Committee's finding that hiding the decline was 'shorthand for the practice of discarding data known to be erroneous'.* However, Russell's finding was set about with declarations in defence of the IPCC consensus – that the divergence problem had been discussed in the literature, that

---

*See p. 132.

the Fourth Assessment Report was not misleading and so on. Of the latter, it was said that despite the spaghetti graph having had the divergence deleted, its depiction of uncertainty was 'quite apparent to the reader'.[296]

REACTIONS

With the Russell inquiry all sewn up, UEA may have felt that they were in the clear, but it soon became clear that there was a certain unease in the media about the integrity of the inquiries. For example, the report was remarkably light on the serious allegations surrounding FOI and this failure was quickly picked up by one of the journalists who examined the draft report. In his coverage of the Russell report, Fred Pearce, the doyen of environmental journalists, pointed out that even the most basic questions had not been asked.

> ...extraordinarily, it emerged during questioning that Russell and his team never asked Jones or his colleagues whether they had actually [deleted emails].[297]

Others also appeared to be uncomfortable, including outlets that UEA might have expected to remain 'on side'. For example *New Scientist* wondered in an editorial how Russell could have concluded that the rigour and honesty of the CRU scientists was not in doubt when there had not been an investigation into whether emails had been deleted. It went on to express concern over the failure of any of the inquiries to look at the science:

> How can we know whether CRU researchers were properly exercising their judgment? Without dipping his toes into the science, how could Russell tell whether they were misusing their power as peer reviewers to reject papers critical of their own research, or keep sceptical research out of reports for the Intergovernmental Panel on Climate Change?[298]

However, these objections were muted, and with environment correspondents in the mainstream media trumpeting the exoneration of Jones and his colleagues, it was clear that UEA were going to win the day. For Acton, Davies and the staff of CRU it looked as if the long struggle was over.

THE PENN STATE INVESTIGATION

In Chapter 6 we saw how Penn State University's examination of the conduct of Michael Mann had been split in two. Just a few weeks after Climategate, an initial inquiry had ruled, in rather murky circumstances, that Mann had no case to answer on most of what they said were the allegations against him. However, on the question of the acceptability of his conduct, they had decided that there were meaningful allegations and so, as required by the university regulations, they passed the question over to a five-man panel of investigation for a determination of Mann's guilt or innocence.

Throughout the spring and summer of 2010, while sceptics attention was focused on the inquiries in the UK, the Penn State panel was working away at its investigation, unnoticed by the majority of observers of the climate scene. Its work barely attracted the attention of sceptics until the publication of its final report, just days before the Russell report was due to appear.[299]

As expected, the Penn State investigation panel's report was not an improvement on the inquiry. Much of the text reiterated the inquiry panel report, sometimes verbatim, but there was one intriguing difference that was quickly noticed: where Easterling had been said by the inquiry panel to have 'recused himself from the inquiry for personal reasons', now there was a slightly different explanation:

> Dean Easterling recused himself from the inquiry due to a conflict of interest.[299]

So extraordinarily, despite having an acknowledged conflict of interest, Easterling still appeared to have had a role in the investigation. He had been asked to identify scientists who might provide evidence to the investigation committee on what exactly were normal standards of behaviour in the field of climatology, a role he undertook alongside an outsider – David Verardo, a senior administrator at the US National Science Foundation. Verardo's was a familar name to McIntyre from the time when he had told the Canadian that Mann's computer code was private intellectual property, a story that directly contradicted the Foundation's published statements on the subject.[42]

As a result of these consultations, the panel had decided to interview two mainstream climatologists – William Curry of Woods Hole and Jerry McManus of Columbia – and the prominent sceptic, Richard Lindzen of MIT. Once again, an inquiry into the behaviour of Hockey Team members was going to fail to interview McIntyre and McKitrick.

With the investigation panel focusing solely on whether Mann's conduct complied with normal standards in the field, the interviews seem to have taken on a somewhat other-worldly feel. The panel's attention focused on peripheral questions, such as whether it was normal to give data and code to other scientists. There seemed to be a concerted effort to avoid addressing any of the actual allegations. Lindzen, in particular, was taken aback at the line of questioning and, apparently unaware of the inquiry panel's findings, demanded to know what had happened to all the other serious allegations that had been made about Mann. His reaction when told that the inquiry panel had dismissed these charges was recorded apparently straight-faced by the investigation panel:

> When told that the first three allegations against Dr. Mann were dismissed at the inquiry stage . . . , Dr. Lindzen's response was: 'It's thoroughly amazing. I mean these are issues that he explicitly stated in the emails. I'm wondering what's going on?'[299]

The panel, it is recorded, 'did not respond to Dr. Lindzen's statement'.[299] Like the other interviewees, Mann was asked about his view of normal behaviour in the field of climatology. However, the committee did go on to ask some follow-up questions, including one on whether he had breached the confidentiality of peer review. Mann's response was that he didn't think he had done so, but he then modified his answer somewhat, saying that he may in fact have done so 'in the belief that permission to do so had been implicit, based on his close collegial relationships with the paper's authors. He suggested that an example of this was the Wahl and Ammann paper, which he said he had forwarded to Briffa.

However, as we have seen, the Climategate emails contained other instances of Mann apparently breaching peer review confidentiality, for example the occasion on which he may have circulated the draft of the McIntyre and McKitrick submission to *Nature* to his colleagues on the Hockey Team.* Even Mann would not have suggested that he enjoyed 'close collegial relationships' with the two Canadians. However, the investigation panel's reading of the emails does not appear to have led

---

*See p. 13.

them to any specific examples on which to question Mann and the conversation moved on to other matters.

Once again the verdict seems to have been a foregone conclusion and Mann was duly cleared of all charges, but it was perhaps inevitable that such a transparently inadequate investigation would not go unnoticed. In particular, the *Financial Times*' Clive Crook wrote a strongly worded commentary on the report, noting that the rationalisation of the verdict given by the investigation committee was quite bizarre.

> The Penn State inquiry exonerating Michael Mann ... would be difficult to parody. Three of four allegations are dismissed out of hand at the outset: the inquiry announces that, for 'lack of credible evidence', it will not even investigate them ... Moving on, the report then says, in effect, that Mann is a distinguished scholar, a successful raiser of research funding, a man admired by his peers – so any allegation of academic impropriety must be false.[300]

And to show he was not exaggerating, Crook quoted directly from the report:

> This level of success in proposing research, and obtaining funding to conduct it, clearly places Dr. Mann among the most respected scientists in his field. Such success would not have been possible had he not met or exceeded the highest standards of his profession for proposing research...
>
> Had Dr. Mann's conduct of his research been outside the range of accepted practices, it would have been impossible for him to receive so many awards and recognitions, which typically involve intense scrutiny from scientists who may or may not agree with his scientific conclusions...
>
> Clearly, Dr. Mann's reporting of his research has been successful and judged to be outstanding by his peers. This would have been impossible had his activities in reporting his work been outside of accepted practices in his field.[299]

However, apart from some expressions of disquiet from a handful of commentators, there was little meaningful questioning of the findings and the media soon moved on to other things. As far as the outside world was concerned climatology and climatologists had been cleared of wrongdoing once again.

As we saw in Chapter 6, as the preliminary inquiry at Penn State had drawn to a close in February 2010, Republicans in the US congress had started to take a close interest in Climategate and there had even been a call for the US government to investigate. Some months later, Senator James Inhofe, a prominent sceptic, wrote to the US Department of Commerce, asking for them to look into the issues that had arisen from the Climategate emails, with particular reference to federal employees. This meant in essence those employed by NOAA, which, rather strangely, is a part of the US Government's Commerce Department. Among the prominent NOAA employees who had worked on the IPCC reports was Eugene Wahl, who had joined the government payroll shortly after his work on the Fourth Assessment Report had been completed. It appears that the Commerce Department felt that it had little choice except to comply with Inhofe's request, and yet another investigation into Climategate was launched.

The Commerce Department inquiry was run along similar lines to the one performed by the House of Commons Science and Technology Committee in the UK, addressing allegations relating to peer review, FOI compliance, data manipulation and integrity of instrumental records. The quality of the investigation of NOAA was, however, much higher than any of the earlier inquiries: some limited attempts were made to corroborate the evidence heard.

The results of the investigation only finally appeared in February 2011 and, like all the inquiries before it, the Commerce Department issued NOAA and its staff a bill of health that, if not clean, was only slightly soiled. However, it was not so much the final conclusions that interested sceptics as one particular observation made by the inspector during his investigation:

> In an email dated May 29, 2008, in which the Director of the CRU requested a researcher from Pennsylvania State University to ask an individual, who is now a NOAA scientist, to delete certain emails related to his participation in the IPCC [Fourth Assessment Report]. This scientist explained to us that he believes he deleted the referenced emails at that time.[301]

The email was of course Jones' infamous 'delete all emails' request, which was sent to Michael Mann along with a request that it be forwarded to Wahl. And if there were any doubt, shortly afterwards a source within the US legislature sent McIntyre some excerpts from the Inspector General's interview of Wahl.

> Q. Did you ever receive a request by either Michael Mann or any others to delete any emails?
> A. I did receive that email. That's the last one on your list here. I did receive that.
> . . .
> Q. So, how did you actually come about receiving that? Did you actually just – he just forward the – Michael Mann – and it was Michael Mann I guess?
> A. Yes
> Q. – That you received the email from?
> A. Correct. . .
> A. To my knowledge, I just received a forward from him.
> . . .
> Q. And what were the actions that you took?
> A. Well, to the best of my recollection, I did delete the emails
> . . .
> Q. So, did you find the request unusual, that they were – that the request – that you were being requested to delete such emails?
> A. Well, I had never received one like it. In that sense, it was unusual.
> . . .
> Q. I guess if the exchange of comments and your review was appropriate, I guess what I'm just trying to understand [is] why you'd be [asked] to delete the emails after the fact, at the time that they're – it appears that the CRU is receiving FOIA requests
> A. Yeah. I had no knowledge of anything like that. But that's what they were – where they were coming from. And so, you'd have to ask Keith Briffa that. I don't know what was in his mind.[302]

So finally, fully eighteen months after Climategate, and after so many investigations and inquiries had failed to establish the truth, sceptics had the answer to one of the most important questions to have arisen from the disclosure of the emails. Mann *had* acted on Jones' request, at least to the extent of forwarding the email to Wahl. And Wahl had done precisely what had been asked of him. It was a sad day for American science, although fortunately for Wahl, the deletion of emails appears to have taken place before he joined NOAA in August 2008.

Confirmation that at least some emails had been deleted in response to Jones' requests was the source of considerable satisfaction to those who had been pressing the question through all those months before-hand. Blogs and social media were soon alight with the news and it was not long before the implications of what Wahl had revealed started to sink in. These were important.

Steven Mosher and Charles the Moderator were first to notice the significance. As they explained in an article published at *Watts up With That?*, if Mann had forwarded Jones' 'delete all emails' request to Wahl then the Penn State inquiry's finding that there was no credible evidence that Mann had 'engaged in, or participated in, directly or indirectly' any attempt to delete emails looked inexplicable. The three members of the inquiry team at Penn State suddenly looked as if they had some very awkward questions to answer. This was potentially devastating for their reputations.

It was not long before the story started to leak into some more main-stream outlets, and in particular *Science* magazine, where journalist Eli Kintisch wrote a long article discussing the furore, although without mak-ing plain the contradiction between the Penn State inquiry report's word-ing and Wahl's new revelations. However he did manage to get hold of of Mann, who was holidaying in Hawaii, and who reacted in characteristic style:

> Mann, reached on vacation in Hawaii, said the stories yesterday were 'libelous' and false. 'They're spreading a lie about me,' he said.... 'This has been known for a year and a half that all I did was forward Phil's e-mail to Eugene.' Asked why he sent the e-mail to his colleague, Mann said, 'I felt Eugene Wahl had to be aware of this e-mail...it could be used against him. I didn't delete any e-mails and nor did I tell Wahl to delete any e-mails.' Why didn't Mann call Wahl to discuss the odd request? 'I was so busy. It's much easier to e-mail somebody. Nowhere did I approve of the instruction to destroy e-mails.'[303]

For all its rage the responses appear to have been carefully worded. Mann appeared to want to give the impression that forwarding the email without comment did not imply approval of the request. Moreover, the question still needed to be asked of whether Mann himself moved or hid his emails when he received Jones' request. But above all, the idea that everyone had known for a year and a half that he had forwarded the email was a bolt from the blue. Not one of the sceptics who had followed the affair so obsessively since the end of 2009 had heard such an admission and none of the inquiries had said so either. In fact, in the immediate aftermath of the Climategate disclosures Mann had given a quite different story, telling one sympathetic journalist:

> This was simply an email that was sent to me, and can in no way be taken to indicate approval of, let alone compliance with, the request. I did not delete any such email correspondences.[304]

In another version of the story, Mann told a newspaper in Pittsburgh that he had alerted another scientist:

> Mann said he did not delete e-mails and regrets that he did not reply to Jones with an e-mail telling him that was an inappropriate request.
>
> 'It put us in an awkward position,' Mann said. Instead, Mann forwarded that e-mail to a colleague to alert him to what Jones wanted the scientists to do.[305]

A MOLE

The Penn State inquiry and the investigation that followed gave every appearance of being another whitewash, but it looked as though there would be nothing for it but to protest the procedural failures in vain, while Mann, Wahl and the university carried on as if nothing had happened. However, someone within the Penn State organisation had other ideas.

Just days after the news about Wahl's deletion of emails had hit the news stands McIntyre wrote a blog post pondering the question of just how much the Penn State inquiry panel had known when they decided that there was insufficient evidence to proceed to a full investigation of the allegations of email deletion. This blog post elicited an almost immediate response from someone within the Penn State organisation who knew exactly what had been going on, but someone who was also very keen to remain anonymous if they were going to tell what they knew – and what they knew would even then only be hinted at and alluded to. The identity of the person concerned is not public, but McIntyre has used the shorthand of 'PS' to refer to them.

PS's first email read as follows:

> You were not contacted for a reason. Recuse but not excuse![306]

That was the totality of the message – there was not even a greeting or signing off. It was very mysterious and to McIntyre it made no sense at all. He quickly sent back a reply:

> I don't understand your comment below. 'Recuse' applies to a judge, not to someone giving evidence. And it wasn't just your inquiry – it was all the inquiries.[306]

To which came the cryptic reply:

> Reread the Inquiry Report for context for 'recuse'.[306]

Still McIntyre didn't understand, and his next reply was somewhat terse.

> I doubt that you would change my mind that your inquiry didn't do its job. Not for the reasons that many readers presume – but because the inquiry purported to assume tasks of the investigation phase that should not be done in an inquiry stage, a point that I made in posts last year ... Procedurally, the process was totally botched. It's too bad that you didn't carry out the job that you were supposed to do, as it might well have somewhat cleared the air.[306]

Once again, PS would only hint at the message he wanted to convey:

> You have reached the right conclusion though based on incorrect assumptions.[306]

It was weeks later when, with the help of some of his contacts, McIntyre finally made sense of what PS was saying: William Easterling, the Dean, who had been said to have 'recused' himself from the inquiry had actually done nothing of the sort. What PS appeared to be saying was that Easterling had continued to run the inquiry in the background, despite having an acknowledged conflict of interest. If this was true then there was no doubt that the inquiry had been a sham.

As if PS's inside story had not been enough, a few months later there was some more evidence to support the idea that Easterling had been running operations behind the scenes. In October 2011, Willam Brune, Mann's immediate boss at Penn State and, as we have seen, a consultant to the inquiry, gave a seminar to his colleagues to outline what had happened. Although this was intended to be a small, virtually private event, several sceptics were on hand to report what was said. One eyewitness recorded a statement by Brune which, although it was not noticed at the time, was later seen to completely confirm what PS had said:

> The first step was an Administrative Inquiry, led by Bill Easterling, the Dean. [307]

So apparently, a consultant to the inquiry was unaware that Easterling had recused himself, and in fact thought that he was running the inquiry. PS's story appeared to be quite true.

CHAPTER 11

THE INQUIRY INTO THE INQUIRIES

SLEIGHT OF HAND

During June 2010, McIntyre travelled to Chicago for the Heartland Conference, an annual convention of climate sceptics from around the world, where he was to be the main attraction. At the start of his keynote lecture he was treated to a standing ovation – a hero's welcome for the man who had done so much to expose the flaws of the paleoclimate papers and to bring the behaviour of the Hockey Team to worldwide notice. To everyone's surprise, however, the reaction to his talk was somewhat muted, after he refused to speak of fraud or demand punishment for the scientists whose failures he had exposed.

One journalist who reported the sceptics' discontent over McIntyre's talk was the BBC's Roger Harrabin, who wrote a somewhat unsympathetic article about the conference, describing the Canadian as 'shambling' and referring to him as 'resembling a character from Garrison Keillor's Lake Wobegon'.[308] Despite being thoroughly unamused by Harrabin's characterisation of him, McIntyre managed to strike up a rapport with the BBC man, who was looking into the stream of revelations that were emerging on the blogs about the way the Oxburgh inquiry had been put together. It emerged that as a result of his newfound interest in a subject that had previously been the preserve only of sceptics, Harrabin had put his own list of questions to Lord Rees, probing the truth of the accusations on the blogs. Rees and the Royal Society were now under pressure from two sides.

The fruits of Harrabin's labours appeared on the eve of the publication of the Russell report in the shape of a ten-minute segment on the *Today* programme, the flagship morning current affairs show on the BBC radio. This was a remarkable change of direction for the BBC, with Harrabin in particular having been regularly criticised by sceptics for the strong 'green' tinge to his journalism and his failure to question mainstream scientists in a meaningful way. To suddenly pick up on a major sceptic talking point and give it a sympathetic hearing on prime-time radio appeared to represent an astonishing volte-face. It almost looked as if the corporation and its key environmental journalist might be about to make amends for their earlier failures.

Although most of the content of the *Today* programme segment covered familiar ground for those who had been following the Climategate story closely, there was one important revelation. This was the news that there was significant concern in the Science and Technology Committee over the stream of revelations about the integrity of the inquiries. Although no longer an MP, Phil Willis, who had chaired the inquiry, was adamant that UEA had misled the committee over the nature of the Oxburgh review:

> Quite frankly I couldn't believe [the Oxburgh report]. I frankly think there has been a sleight of hand in that the actual terms of reference were not what we were led to believe.[309]

Harrabin had tried to get an interview with Oxburgh, but had only managed to extract a written statement in which Oxburgh had brushed aside the concerns about his panel's work. He said that the panel had read much more widely than just the short list of papers with which they had been supplied and he also claimed that the university had only ever asked him to assess the integrity of the scientists rather than the quality of the science. As Harrabin noted, this story did leave much to be desired, since there were no records of which papers the Oxburgh inquiry had actually examined. It also left open the question of how it was that the Science and Technology Committee had been led to the incorrect conclusion that Oxburgh would look at the science.

Certainly, this mistaken idea had spread well beyond the confines of the Science and Technology committee. By July the UK's general election was over and Parliament was up and running again. A new Science and Technology Committee was selected, and its members promptly began a scene-setting inquiry to help them come to terms with their new roles. Among the witnesses called were Lord Rees and the newly installed Science Minister David Willetts.

Rees was certainly keen to leave the committee with the impression that all was well with CRU's science, disputing an assertion from Stringer that none of the inquiries had examined the science:

> I would, to some extent, contest what you have just said ... the key thing which the Oxburgh Committee did was to actually go and sit with the scientists and see what they actually did and how they analysed the data...I do not think the science from [CRU] is severely under question from the techniques they used.[310]

And Willetts was pushing the same line:

> ... although there are lessons to be learned I think they show that, when it comes to the conduct of the science, the work that was done at UEA as I understand it, has passed muster when assessed by independent experts to check whether anything went wrong. My view is that their scientific work stands.[311]

It is particularly interesting to speculate how Willetts arrived at such a mistaken understanding of the meaning of the Oxburgh report. It is, of course, possible that he read about it in the newspapers, but it is more likely that he was briefed by civil servants in his department: either by his departmental scientific adviser or, more intriguingly, by the government chief scientific adviser, Sir John Beddington, who works in the same department as Willetts but reports directly to the Prime Minister.*

THE GUARDIAN DEBATE

Although government and Parliament were getting a highly dubious explanation of the work undertaken by the inquiries, there were soon to be opportunities for sceptics to put their side of the story and another important chapter in the Climategate story was to be opened.

At the end of June, McIntyre had been contacted by the environmental journalist Fred Pearce, who had been spearheading the *Guardian*'s coverage of Climategate and had just published a history of the affair.[94] Pearce wanted to know if McIntyre would be willing to take part in a panel discussion on Climategate and the findings of the inquiries – this was scheduled to take place in London shortly after the publication of the Russell review in early July, and would feature Trevor Davies, Doug Keenan and Bob Watson. Unfortunately, shortly after Pearce's approach, McIntyre was contacted by the *Guardian* who said that they would have to withdraw the invitation because of cost constraints. However, a collection was quickly organised by the sceptic blogs and shortly afterwards McIntyre headed to London.

McIntyre's visit generated enormous interest among UK sceptics – it appeared that he was finally going to get his chance to set the truth out in public. There was, however, a great deal of work to be done to get ready.

---

*Beddington runs the Government Office for Science, which is part of the Department of Business, Innovation and Skills. Willetts is a junior minister in the department, with responsibility for science.

There was only a week between the publication of the Russell report and the date of the debate, and so McIntyre was going to have just these few days into which to compile an analysis of the Russell report and the background documents that had been published at the same time. On top of this burden, a series of meetings and other engagements had been set up with people keen to meet the man who had broken the Hockey Stick. Getting himself ready for the debate was going to be no easy task.

The debate was held in the imposing surroundings of the Royal Institute of British Architects in central London.* The format was unremarkable – each of the speakers had five minutes to make an introductory statement, after which the floor would be opened to questioners. Proceedings were overseen by the prominent environmental journalist George Monbiot, who might at first have been seen as a less than neutral chairman but in the event was generally seen as having managed proceedings in an even-handed manner.

The opening speaker was Trevor Davies of UEA, who, perhaps predictably, wanted to discussion the series of exonerations delivered up by the inquiries:

> [T]he revelations [of the emails], as interpreted by some on the basis of a very small sample from a great number over a long period, were far from reality. I say this on the basis now of three independent investigations into UEA...[†][312]

However, despite having the headlines from the Oxburgh and Russell reports to hide behind, Davies was clearly still keen to avoid any discussion of the failings of the inquiries and devoted much of his time at the lectern to explaining all the lessons that had been learnt at UEA – the need for openness and for clear explanations of scientific uncertainty for example. The message from the university was therefore plain: it was time to move on.

As Monbiot introduced each of the speakers, he asked each of them an opening question, to which they responded before launching into their allotted five minute presentation. When it came to McIntyre's turn to speak, Monbiot's question was whether he had been wasting everyone's time by spending so long trying to obtain the CRUTEM surface temperature data. Monbiot appeared to view this as a penetrating opener, apparently

---

*The story of the debate is based on an unofficial transcript.[312]

[†]Davies also mentioned the Penn State Investigation. See Chapters 6 and 10.

not realising just how peripheral CRUTEM was to McIntyre's work and to the Climategate affair as a whole, and was somewhat taken aback when McIntyre flatly rejected the idea that he had dedicated himself to this area of CRU's work. Despite this Monbiot returned to the subject later in the debate, asking McIntyre again whether he had been wasting everybody's time in seeking CRU's data and code since Russell had said he had been able to broadly replicate Jones' work from publicly accessible data. It is perhaps rather surprising that such a prominent environmental journalist, and one who had written widely on Climategate, had failed to grasp what McIntyre's work was about, although the fact that the Science and Technology Committee and the Russell panel had both decided to focus their attentions on CRUTEM could perhaps mislead the unwary. The Science and Technology Committee's reasons for looking at CRUTEM were a mystery (and remain so) but, as we have seen, apparently unwilling to mention Beddington's role in directing his panel's work, Russell had claimed that criticism of CRUTEM was 'implied', however unconvincing a story this might have been.*

Having dealt with Monbiot's red herring, McIntyre delivered a succinct summary of the failings of the Climategate inquiries – the misrepresentation of Oxburgh's terms of reference, the failure of Russell to attend the interview of Jones, the failure to interview critics and the inadequate investigation of serious allegations such as the 'delete all emails' request and the hide the decline episode. He concluded by noting what the effect of the failure to address the problems in climate science would be:

> If climate scientists are unoffended by the failure to disclose adverse data, unoffended by the 'trick' and not committed to the principles of full, true, plain disclosure, the public will react, as they have, by placing less reliance on the pronouncements from the entire field.[312]

Bob Watson, the former head of the IPCC who had been so prominent in the media when Climategate broke, was the next to speak, covering much of the same ground as Monbiot and Davies – he spoke about the integrity of the reviews and, like the inquiries themselves, cited the reliability of the CRUTEM surface temperature record. However, the impact of his introductory remarks was somewhat blunted by his admission that he

---

*See p. 203.

had read very few of the emails anyway,* a declaration that caused hoots of derision from the audience.

Keenan, meanwhile, started off by noting that he had accused Jones of fraud, explaining that neither the Russell nor the Oxburgh panels had seen fit to consider his allegations, despite the story having been splashed across the front page of the *Guardian*. He also covered more general questions of the integrity of science: the problem, he said was that there was no system of accountability to ensure scientists' good behaviour.

> Both the Russell review and the Oxburgh review were clearly white-washes. But that is not the problem. The real problem is the lack of systemic accountability. There should be some general mechanism in place whereby allegations of improper behaviour are dealt with. What kind of society would we have if there were no police, judiciary or prisons? That, in effect, is the system in place in science today. There are tens of thousands of scientists in the United Kingdom. As far as I know, none have been convicted of research fraud in the last twenty years. That is not credible. Even among much smaller groups of respected people, for example members of Parliament, Catholic priests, police detectives, frauds do occur. [312]

Fred Pearce's contribution was keenly awaited; as the doyen of environmental journalists in the UK he was respected by everyone on the green side of the debate, but his book on the Climategate affair had revealed him as someone who was willing to follow the facts rather than any ideological agenda. [94] His presentation was very much in this vein, opening with an assessment of the inquiries that was in stark contrast to that of Davies a few minutes earlier:

> None of the reviews really explored the science. They were concerned principally with process, and I think there are quite a number of issues where... they simply... said, 'Well, this is the scientific judgement of the scientists concerned, we have to accept that, because we're not delving into the science...' [312]

Pearce's views on the Climategate affair as a whole were also important, having much in common with the views of many sceptics:

> I was... disturbed by the emails I think when I first read them, I quite early on thought that they merited investigation, and I'm still

---

*With the benefit of hindsight it is strange to notice the parallel to the Penn State inquiry, and Gerald North's admission that he too had not read the emails. See p. 120.

quite disturbed about the emails and what they reveal about the conduct of some of the scientists involved – not just, I might say, the CRU scientists. But for me, I think, Climategate is not a conspiracy, it's a tragedy really, it's a tragedy born of mistaken judgement of motive. It seems to me that after years of fighting off commercially and politically motivated critics, the scientists at CRU and their colleagues...lost sight of the fact that there might sometimes be value in those criticisms. They just saw them as criticisms from enemies and left it at that. They failed to spot that there was also a new generation of critics – we have two of them on the platform here – who are essentially I think data libertarians rather than climate sceptics, still less climate deniers. So the mainstream scientists, they responded to criticism by shutting up shop, by refusing to share data, adopting if you like a kind of siege mentality, and that's what for me came again and again and again through the emails, that kind of siege mentality.[312]

Pearce's words could have been taken straight from the pages of the book that Mosher wrote with journalist Tom Fuller immediately after Climategate.[313]

However, Pearce wasn't putting all the blame on one side, telling the audience that sceptics had interpreted the siege mentality at CRU as evidence of the Jones and his colleagues hiding some scandalous secret. While this might have been true of some less-informed commentators, McIntyre in particular had said repeatedly that he thought CRU's stonewalling of FOI requests was only a case of their trying to draw a veil over the routine nature of the work they did on CRUTEM and the minimal quality control measures they had put in place. He had specifically ruled out the idea that they were tampering with the data.[314]

Pearce also summarised his feelings about Climategate in a way that few sceptics would disagree with:

So, as I say, there's no grand conspiracy, but there was some rather grubby behaviour going on among the scientists emailing each other. Bob [Watson] has kind of listed the charges against them, and I think they're difficult to dismiss, about abuse of the spirit and sometimes of the letter of the Freedom of Information law, about conflict of interest in peer review, about abuse of the rules supposed to require open review at IPCC. And there were abuses that ended up with, you know, round robins asking for emails to be deleted, and you know that is not acceptable behaviour, I don't think. And I don't think, as I said, I don't think any of the official reviews really got to grips with that.[312]

Although Pearce went on to note that the UK inquiries were better than the Penn State inquiry,* as his remarks here make clear he felt that many allegations, including some of the most serious ones, had not been addressed by the inquiries. Once again, sceptics in the audience found themselves in complete agreement with the environmental journalist.

Later in the evening, during the question and answer session, Monbiot returned to several of these themes, in particular asking Davies about the inquiries. Davies insisted that both Oxburgh and Russell had been completely independent of the university and, equally surprisingly, seemed to want to dispute the idea that the science had not been addressed:

> I was a little puzzled about the comments that the science wasn't addressed. When... even in the Muir Russell review there was quite a deep examination of some aspects of the science, especially in the way that CRU responded to the Muir review questions. Its first submission to the Muir Russell panel which was put on the Muir Russell website in March or April, [was] seventy or eighty pages, mostly quite detailed science.

To Davies' way of thinking, Oxburgh's remit – 'to see whether data had been dishonestly selected, manipulated and/or presented to arrive at predetermined conclusions that were not compatible with a fair interpretation of the original data' – *was* a review of the science and in fact was completely compatible with the statements made to Parliament. Unfortunately, this suggestion directly contradicted Davies' earlier admission to Beddington that Russell would *not* be examining the science, as well as Oxburgh's blunt statement that the science had not been the subject of the inquiry.[†]

As the discussion moved on, the sceptics in the audience were starting to make their presence felt. Keenan in particular was winning regular applause with his denunciations of the quality of climate science and the lack of transparency and accountability among climatologists. Perhaps the most remarkable moment was when McIntyre explained that Russell had failed to attend his own inquiry's interview of Jones, leaving that task to other members of the panel. When Monbiot invited Davies to dispute this, the UEA man first obfuscated, was left temporarily speechless, and was then reduced to asking McIntyre to help him with some of the details

---

*See Chapters 6 and 10.
[†]See p. 147 and p. 162 respectively.

of the inquiry, much to the amusement of the audience. In many ways the ridicule seemed to epitomise the Climategate inquiries.

THE GWPF REPORT

The second opportunity to sway public opinion came two months later, with the publication of a report on the inquiries themselves, and here, as the author of the report, I must shift the narrative to the first person.

On the day of the publication of the Oxburgh report I had been approached by Benny Peiser of GWPF with a view to writing an 'inquiry into the inquiries', an invitation I readily accepted. The questionable performance of the Science and Technology Committee inquiry had already produced a great deal of material, as had the obvious bias in the composition of the Russell and Oxburgh panels. However, over the following weeks the steady stream of revelations of the way the public had been misled and the outcome of the inquiries influenced meant that it had become possible to put together an unassailable case. The GWPF report promised to produce fireworks.

Over the start of the summer, I spent many hours working away at the report, squeezing it between my day job and the demands of my blog and family. However, at the start of June I found myself in London, Peiser and I having been invited to a lecture to be given by Rees at the BBC. By coincidence, or perhaps more likely by design, among the other invitees was none other than UEA's Trevor Davies, and after the lecture he approached us for a chat.

The three of us spoke cordially about the way the inquiries were moving, but little of any consequence was said. However, the following day Peiser met Davies again at a discussion meeting on the subject of global warming policy in the wake of the failed Copenhagen conference – the speakers were Oxburgh, Beddington and one of Oxburgh's colleagues in GLOBE, Lord Jay.* During their conversation, Peiser decided that as a courtesy he should reveal GWPF's intention to conduct an 'inquiry into the inquiries'. So, long before the report itself appeared, the authorities at UEA and in the scientific establishment were aware that trouble was coming, although they had no idea of precisely when it would hit them – even I was in the dark, since Peiser had yet to set a publication date. Nevertheless, the university and their supporters could still take steps

---

*GLOBE was discussed at p. 148. The discussion meeting was held at the Foundation for Science and Technology, see www.foundation.org.uk.

to minimise the damage: Davies quickly passed word of my report onto Beddington.[315]

The formal announcement of the GWPF investigation took place a month later on 7 July 2010, coinciding with the publication of Russell's findings and Harrabin's *Today* programme interview with Willis. The press release was brief and to the point, but Davies and the Hockey Team would no doubt have been pleased to see that it included the one detail they needed most: the publication of the report was slated for the end of August. Time was therefore short – the storm could be breaking upon them in as little as six weeks – but it appears that preparations were well in hand.

The first hint that a great deal of effort had been expended over the summer months came when I was telephoned by Ben Webster, the environment editor of the *Times*, who surprisingly wanted to know if my blogging work was received financial support from outside bodies. At that time I occasionally put a tip box up on my blog to allow readers to contribute if they wished, and Webster pressed me for details of how much I was receiving, whether there were any big donations and whether any of the donors appeared to come from industry. I decided to adopt an approach of complete openness, and was able to set his mind at rest on the subject – the sums involved were paltry and all appeared to be from private individuals. However, despite my making such a full disclosure, Webster appears to have decided to publish a story regardless.

The *Times* of 19 July 2010 was a remarkable edition of this most venerable of newspapers. The front page was emblazoned with a Webster article about Exxon-Mobil having funded think tanks that were involved with climate scepticism, their largesse amounting to some £1m over the previous year. This was clearly a 'hit piece', with no news value at all. It was notable that Webster made no mention of the fact that this money was actually funding the think tanks themselves, rather than their global warming activities: much of the money was therefore being spent in other policy areas entirely that had nothing to do with global warming at all.

The article continued on page 5, with Peiser and GWPF put in the frame by the bizarre means of noting Peiser's *attendance at a conference* – this had been organised by one of the Exxon-funded think tanks.* On

---

*The conference predated the existence of GWPF and so Webster's point is somewhat specious.

the same page another Webster article looked at Climategate and said that UEA scientists were still being 'hounded' by sceptic bloggers whose attacks, Webster said, could be 'dismissed as the rantings of mildly obsessive individuals', although he did at least note that there was no evidence that CRU's critics were being funded by oil companies. And if that were not enough, page 2 carried a leader article that sang the praises of the IPCC's 'unanswerable case' and rammed home the message about the links between sceptics and big oil.

Soon afterwards, the temperature was raised even further with what appears to have been a concerted campaign to discredit me. Until that time it was widely assumed that green activists had adopted a strategy of ignoring *The Hockey Stick Illusion* entirely rather than giving it any publicity: in the first six months after the book's publication barely a single critical review had appeared.* However, it appears that the advent of the GWPF report prompted a change in tactics and from mid-July a series of fiercely critical blasts against the book – or more often against me as its author – appeared. Nick Hewitt, an atmospheric chemist and aspiring IPCC author, described the book as 'tedious' and 'pedantic', although slyly admitting to the existence of 'shortcomings in working practices' among paleoclimatologists.[317] Another chemist, Richard Joyner, referred to the book's 'wickedness' and said it was 'a McCarthyite book that uses the full range of smear tactics to peddle climate change denial'.[318] The first review that tried to contest some of the science presented in *The Hockey Stick Illusion* then appeared at *RealClimate* under the byline of the pseudonymous author, Tamino,[319] but he was soon shown to have relied on some extraordinarily cynical quoting out of context to achieve his effect.[320]

It was clear that this was not going to be the end of the attempts to discredit me before the report appeared. Bob Ward had been trying to extract the exact publication date of the GWPF report for some time, but had been brushed aside by Peiser, who had refused to reveal any details until he was ready to do so. Ward was clearly up to something and was trying to time his efforts for maximum effect. However, thwarted by Peiser he had to make a guess and unfortunately he guessed incorrectly – the report had been delayed beyond the originally intended date, due principally to the steady stream of new revelations that needed to be

---

*The lack of critical reviews was proving a problematic for the book's Wikipedia page where it was felt that a critique would provide balance.[316]

incorporated into the text. Ward's article – yet another critical review of the *The Hockey Stick Illusion*, this time in the *Guardian*, ended up preceding the report by three weeks. This was an astonishing attack,[321] which accused me of having 'a history of omitting evidence' to suit my purposes and contained a series of more or less fabricated accusations. For example, Ward claimed that I had omitted important facts about Hans von Storch's resignation from *Climate Research*:

> ...nowhere does Montford find space for von Storch's own explanation...that he had resigned 'to make public that the publication of the Soon & Baliunas article was an error' because it suffered from 'severe methodological flaws'.[321]

This was a surprising thing to say, because when von Storch is introduced as a character in the *The Hockey Stick Illusion*, he is described as follows:

> Von Storch is one of the big names in climatology and had been one of the editors who had resigned from the board of *Climate Research* over its publication of the Soon and Baliunas paper...[322]

By the time the GWPF report appeared on 14 August, the *Guardian* had issued a partial correction and an apology. So although Ward's false statement about von Storch remains uncorrected, the impact of his poison penmanship was largely lost.

Quite apart from the media onslaught, there were some tricky decisions to be taken. Shortly after the publication of the Russell report, Graham Stringer had written to the panel asking how the recommendations of the Science and Technology Committee had been implemented and in due course Russell had published his response on the inquiry website.[323] Of itself, this correspondence appeared relatively unimportant, but it did suggest that at least one member of Parliament was still interested in the question of the integrity of the inquiries. Then, at the start of September and with the press conference for the GWPF report just days away, the House of Commons suddenly issued a notice of its own 'inquiry into the inquiries'.

> The Science and Technology Committee will hold an oral evidence session following-up to the previous committee's report on the disclosure of climate data from the Climatic Research Unit at the University of East Anglia...

> The Committee will take evidence from Lord Oxburgh, who headed the International Panel that was set up by the University to assess the integrity of the research published by the Climatic Research Unit.
>
> An oral evidence session with Sir Muir Russell, who headed the Independent Climate Change E-mails Review, will be announced in October.
>
> The sessions will focus on how the two reviews responded to the former committee's recommendations about the reviews and how they carried out their work.[324]

The announcement was welcome, but created something of a problem: the committee were to hear from Oxburgh on 8 September and Russell a month later, which meant that the GWPF report was going to fall right in the middle of the two hearings. For some days we deliberated over whether to put the date back until both inquiry chairmen had been heard as this would allow us to challenge anything they told the committee. However, with the date for the report already announced it was decided to push on regardless and the GWPF team started to make preparations for the press conference, by then just two weeks away.

While sceptics were pleased with the news that the committee would be questioning Russell and Oxburgh, with three whitewashes already delivered, there was little expectation that this new inquiry would be much different to the others, particularly as the principal critics of CRU and the inquiries would once again not be given the opportunity to make their case. As one blog commenter put it:

> At most, I'm expecting a couple of hard-sounding questions to be answered with half-assed responses, and the potential hard-line of questioning to just...fizzle.

This turned out to be a very prescient observation.

THE OXBURGH HEARING

The new Science and Technology Committee was very different to the one that had looked into the Climategate affair a few months earlier (see Table 11.1). Most of the members of the previous committee had moved on following the parliamentary elections in May 2010 – standing down like Phil Willis or voted out of office like Evan Harris. Graham Stringer was the only familiar face and also almost the only member of the new

TABLE 11.1: Members of the Science and Technology Committee at the time of the second Climategate hearings

Name	Party
Andrew Miller (Chair)	Labour
Gavin Barwell	Conservative
Gregg McClymont	Labour
Stephen Metcalfe	Conservative
David Morris	Conservative
Stephen Mosley	Conservative
Pamela Nash	Labour
Jonathan Reynolds	Labour/Co-operative
Alok Sharma	Conservative
Graham Stringer	Labour
Roger Williams	Liberal Democrat

Source: STCE II.

committee with a significant scientific background, although since the inquiry looked as though it was going to address procedural rather than scientific matters this was perhaps not the concern it might have been. The new committee was led by Andrew Miller, a genial greybeard whose earlier career had spanned periods as a laboratory technician and as a trade union official.

Miller opened proceedings by asking Oxburgh to explain how the panel's membership had been selected.* The response was long but skirted around the truth, with Beddington's role not mentioned and with a long and misleading description given about some of the panellists. For example, Graumlich was described in the following terms:

> Lisa Graumlich...was a tree ring person, but had not used tree rings in the same way as the Climatic Research Unit used them. Basically, the traditional use of tree rings is for setting up chronologies – you know, the archaeological applications. What the Climatic Research Unit did was, of course, try and interpret the characteristics of these rings in terms of climate, and that is a different step and quite a difficult step to take.[325]

This was a disturbing start to the hearings, since Oxburgh's characterisation of Graumlich's work had little basis in fact. In his post-mortem on

---

*See note on official transcript at the start of the bibliography.

the hearing, McIntyre pointed out that the truth was that she had written dozens of articles interpreting tree rings in terms of climate. Indeed, just a few weeks after the publication of the Oxburgh report, she had given evidence to the US House of Representatives on that very subject.

McIntyre also pointed out that what the committee had heard from Oxburgh about one of the other panel members was equally untrue:

> ...to say or imply that Emanuel was 'outside the [climate] field' or had not taken a position on climate issues is obviously untrue. Emanuel has taken a strong and very public position on climate change. And while Emanuel had not collaborated with CRU itself, he had, of course, collaborated with their closest Climategate correspondent, Michael Mann.[326]

Some of the questions asked also seemed to misrepresent what the actual criticisms were. The fact that Michael Kelly was somewhat sceptical of global warming had been clear for several months by the time of the hearing but despite this, Miller asked Oxburgh why there had been no sceptic on his panel. This allowed Oxburgh simply to point out that one of his panel members *was* a sceptic. And with Miller and the other committee members failing to follow up with any questions about the balance of the rest of the panel's membership or the privately declared intention to have only *some* members who would look at the inquiry with 'questioning objectivity', a serious failing of the inquiry was neatly sidestepped. It already looked as if the second Science and Technology Committee inquiry was going to be little better than the first.

The terms of reference

After their brief look at the panel membership, the committee turned to Oxburgh's terms of reference and here there was a brief moment of clarity, when Oxburgh stated that he thought that Acton had been 'inaccurate' when he told the first parliamentary inquiry that Oxburgh's panel was going to assess the science.[327] He said that his intention to look only for evidence of misconduct had been clear in the press release that announced the panel but, somewhat eccentrically, excused Acton's having reported something rather different to Parliament on the grounds of his inexperience in the vice-chancellor's role. Miller's line of questioning was picked up by another member of the panel, Stephen Mosley, who said that the 'people who had been kicking up a fuss' had said that the terms of reference had been changed. This again was not correct: the allegation was that the public and parliament had been told one thing while the panel had investigated another. This is what sceptics had alleged and what Phil Willis had spoken of in his interview with Roger Harrabin,* although the BBC man had then characterised Willis's concerns as being about a 'shift' in the remit of the panel. In response to Harrabin's interview, UEA had contrived to issue a denial framed in the terms of Harrabin's inaccurate summary of Willis's complaint rather than the actual allegation: they said, no doubt quite correctly, that they had not asked Oxburgh to 'adopt a "narrower brief" of any kind' subsequent to his appointment.[328] On the question of whether they had misled Parliament about the terms of reference, however, the university had maintained a diplomatic silence.

Mosley may have been badly briefed on the details of the allegations – a remarkable thing given Willis's very public airing of them – but he was at least keen to push the point, managing to elicit the information that the decision to focus on misconduct had been agreed in a conversation between UEA staff and Oxburgh when the university were trying to persuade him to take on the role of inquiry chairman. Since Oxburgh was known to have been working on the inquiry by the end of February 2010,† the finalisation of the terms of reference must have taken place before then. In other words Oxburgh had decided not to assess the science *before* Acton had told the original Science and Technology Committee hearing – on 1 March – that the inquiry would assess 'the whole of the science'.

As an aside, we should also question why, if he was already working on the inquiry at the time, Oxburgh's name was not mentioned at the original Science and Technology Committee hearings. An obvious explanation presents itself: with Philip Campbell's forced resignation from the Russell panel still fresh in their minds, the UEA team would presumably have been keen to avoid a repeat of such a public humiliation. It is therefore quite likely that they decided to hold back Oxburgh's name for a few weeks so that they would not be placed in the uncomfortable position of having to explain to Parliament why they had appointed a chairman with conflicts of interest.

---

*See quote on p. 220.

†The correspondence cited on p. 148, which is the first evidence of Oxburgh working in his role as chairman of the inquiry, is dated 27 February 2010.

Due diligence

One of the Conservatives on the committee, Alok Sharma, correctly noted that the inquiry had been criticised for the speed at which it had completed its work, suggesting that this had taken 'only three weeks'. At this point, Oxburgh half-corrected him, explaining that the work had been performed 'over four days, five days – something like that'. But in fact even this was rather misleading: amongst the email correspondence obtained from David Hand* was a detailed itinerary for the panel's visit to the university that showed that the time spent interviewing the CRU scientists was restricted to just two meetings totalling not much more than two hours (see Table 11.2). Although Graumlich and Hand had made a separate visit, the total time spent on the investigation was clearly inadequate to determine if anything untoward had taken place. However, Sharma and his fellow committee members appeared satisfied by an assurance that 'immense amount of work' had been done beforehand and the subject was dropped, without even an inquiry as to what meaningful work could have been achieved offsite beyond reading the original papers.

Keenan and the China stations

It was Graham Stringer who raised the awkward question of the fraud accusation that Keenan had made against Phil Jones. Although the list of papers considered by Oxburgh and his team contained few that were controversial, the one notable exception was Jones' UHI paper and it was therefore remarkable that the Oxburgh report had not even mentioned Keenan's allegations.[70] The exchange between Stringer and Oxburgh was simply astonishing:

> STRINGER: One of the accusations made in the evidence to the predecessor committee of this was made by Keenan. . .
>
> OXBURGH: Made by?
>
> STRINGER: Keenan, who accused Professor Jones of fraud.
>
> OXBURGH: Yes.
>
> STRINGER: If you were trying to find out whether there was fraud going on or whether the scientists had integrity, did you look at Keenan's accusations?
>
> OXBURGH: I don't recall doing so, if I did.[329]

---

*See p. 162.

TABLE 11.2: Itinerary for the Oxburgh panel's visit to CRU

Wednesday 7 April

9:30 am–9.45 am	Taxi to CRU ... Met by Acting Director, CRU, Prof Peter Liss... Coffee and tour round CRU.
9.45 am–10.45 am	Meeting with Phil Jones, Tim Osborn and team in CRU library 30 minute presentation by Phil Jones followed by questions.
10.45 am–11.00 am	Coffee served in CRU library.
11.00 am–12:30 pm	Discussion – CRU library.
12:30 pm–1:30 pm	LUNCH for panel members ... [at] CRU.
1:30 pm–3:30 pm	Discussion – CRU library.
3.30 pm–4.30 pm	If needed: follow-up meeting with Phil Jones and Peter Liss.
4.30 pm–5.30 pm	Panel private meeting
5.30 pm	Peter Liss to chaperone panel ... for taxis to hotel.
7.00 pm	Working dinner at Caistor Hall.

Thursday 8 April

8.45am–9.00 am	Taxi to CRU... Met by Acting Director, CRU, Prof Peter Liss. Coffee in CRU.
9.15 am–10.45 am	Meeting with Phil Jones, Tim Osborn and team in CRU library.
10.45 am–11.00 am	Coffee served in CRU library.
11.00 am–12:30 pm	Discussion – CRU library.
12:30 pm–1:30 pm	LUNCH for panel members...
1.30 pm–3.00 pm	Final meeting.
3.00 pm–3.30 pm	Coffee + depart in taxis...

Source: Hand emails.

So according to Oxburgh's evidence, he not only had failed to investigate Keenan's allegation but in fact he had *not even been aware of it*, despite the accusation having been set out in writing in evidence to the Science and Technology Committee.[186] Oxburgh's ignorance of one of the most serious allegations made against CRU scientists is even more extraordinary when it is recalled that the *Guardian* had reported Keenan's allegations on its front page, declaring that Jones and his colleagues had 'tried to hide flaws in a key study'.[223] There can be little doubt therefore that Oxburgh should have known about Keenan's allegations. His ignorance appears then to be conclusive evidence that his was not a serious attempt to get to the bottom of the Climategate affair.

The list of papers and the raw materials

Stringer's questioning also extracted an admission from Oxburgh that the list of papers had been selected by the university rather than by the Royal Society, as UEA and Oxburgh's own report had led the public to believe. In fact, a week after the questioning of Oxburgh, McIntyre had written to UEA and had elicited an admission that Trevor Davies had been responsible for compiling the list and for appending to it the false claim that the papers had been chosen because of their 'pertinence to the specific criticisms...levelled against CRU's research findings'.[330] This appeared to demonstrate clearly that the Oxburgh report had been misleading, although Oxburgh himself was keen to point out that his team had used the list only as a starting point for its inquiries.

> Let me emphasise, [these papers] were just the start. Because all of us were novices in this area, I think we all felt that they gave us a very good introduction. From then we moved on. We looked at other publications. We asked for raw materials and things of that kind. The press seems to have made quite a meal of the choice of publications. I think for anyone on the panel this all seems a bit over the top because it didn't have that significance.[329]

Given that the Oxburgh panel had completed their work at UEA in a matter of a few hours, the suggestion that they had examined raw materials stretches credibility to breaking point. It is quite certain that little meaningful examination of either further papers or raw materials was possible in the timescales involved.

Stringer continued to press Oxburgh on the theme of the raw materials by inquiring if the panel had determined whether CRU were able to recreate their original results from original data.

> STRINGER: When the panel was carrying out its appraisal were the scientists at CRU able to make accurate reconstructions from the publication back to the raw data that they themselves had used?
>
> OXBURGH: Not in every case. Not with the old material.

> STRINGER: You have surprised me over a number of things you have said, Lord Oxburgh. That is very surprising, isn't it?
>
> OXBURGH: I think it's undesirable but it isn't too surprising.[331]

Oxburgh went on to explain that in his view it was perfectly normal practice in universities to throw out completed work and to fail to document it properly, and he contrasted this with the norms in the commercial world. In fact, as the questioning went on Oxburgh made a series of important revelations that, while not directly pertinent to an investigation of CRU scientists' honesty, were completely damning of the standards of science at CRU – such an assessment was of course precisely what Parliament and public had been led to believe Oxburgh would deliver. For example, back in July, McIntyre had reported that Jones had told the Oxburgh panel that it was 'probably impossible to do the 1000-year temperature reconstructions with any accuracy'.*[231] When Stringer asked Oxburgh about this claim, Oxburgh denied that Jones had said any such thing, but said that had he done so it would have been true. Oxburgh also found little to disagree with when another committee member suggested that CRU scientists were out of their depth on statistical matters, although he suggested that similar results would have been obtained even if they had have used appropriate methods.† To say the least, this made science minister Willetts' declaration that CRU's science had been given a clean bill of health look even more preposterous than it did already.

The Kelly paper

One last element of the questioning bears examination, because it was going to have important implications on what was to follow. Stringer told Oxburgh that he had been sent a copy of the Kelly paper, and asked whether it would not have been better for the inquiry to have discussed it in the report. Oxburgh attempted to shrug off the questioning:

> Again, we were working to time...Look, there are many ways of ‚skinning a cat. Someone else doing this might have done it quite differently, but I am content with the way that we did it and I don't think I would do it differently again.[332]

---

*Although unidentified, McIntyre describes his source as 'reliable'.

†Readers of the *The Hockey Stick Illusion* will be familiar with the extraordinary argument that the use of inappropriate methods does not prevent climatologists from reaching the correct answer.

However, Stringer was insistent, quoting several of the most critical sections:

> STRINGER: 'This is turning centuries of science on its head'. There are a lot of comments like that. 'It is hard directly to correlate this aspect with the anthropogenic hypothesis of climate warming. Some features do correlate – others don't – so where [are] the rigorous tests of the significance of correlation or lack of it?' One could go on.[329]

Oxburgh again insisted that there was nothing to Kelly's comments, explaining that people were reading too much into what were just the comments of a physical scientist looking at the work of observational scientists. To everyone watching, this must have appeared to be a very weak answer and one that did little to deflect Kelly's criticisms of CRU or enhance Oxburgh's reputation. Perhaps then, it was inevitable that this was not the end of the story of the Kelly paper.

Oxburgh's secret evidence

On completion of an oral hearing in Parliament, committee staff circulate copies of the transcript to those involved, giving witnesses the chance to correct or clarify what they have said. However, Oxburgh went much further than merely making corrections, producing a completely new new explanation of the Kelly paper.*

In his new story, Oxburgh made the remarkable assertion that Kelly had been commenting on climate science as a whole rather than CRU's work in particular:

> Kelly's observations. . . could, taken out of context, have been very misleading. They could be taken as a serious criticism directed at CRU when they are in fact a comment on the language and practice in climate science as a whole. They had no bearing on our inquiry into the scientific integrity of CRU.[329]

This was an extraordinary claim. Kelly's paper is structured in two halves – first his comments on Briffa's papers and then those for Jones' work. There is therefore no possible way to construe his comments as

---

*It is not clear precisely when Oxburgh returned his transcript. The Clerk to the Committee has said: 'I am afraid that we no longer have a note of exactly when the transcript was returned but, at the very latest, it had been received by 5 October 2010, when the comments were added to the final corrected version of the transcript.'[333]

being anything other than direct criticism of the science conducted at CRU. In fact, both the criticisms that Stringer had highlighted during the oral hearings could both be tied to specific Briffa papers.* Moreover, Kelly is not a climatologist, so could hardly be expected to have any knowledge of 'climate science as a whole'.† Finally, if there were any lingering doubt that what Oxburgh said about the Kelly paper was wrong, Kelly himself has now quashed it.[334]

Oxburgh's remarks about the nature of Kelly's critique were transparently incorrect and they would have been challenged immediately had they been seen by any of the sceptics who were monitoring all the changes to the committee's website. It is therefore of some concern that Oxburgh's new story did not appear in public before the Science and Technology Committee's new report was published in January: for nearly four months, its existence was known only to Oxburgh and committee insiders.

THE GWPF REPORT

The excitement over the revelations Oxburgh had made during the hearing had barely died down before it was whipped up again by the publication of my report on the inquiries. Although most of the significant issues had already been made public on my blog, there had been little reaction from a largely uninterested mainstream media, with the notable exception of Harrabin's radio piece at the time of the publication of the Russell report. The GWPF report was therefore a good opportunity to force the issues into the headlines. What was really needed, however, was some new piece of information that would give the journalists a good headline to get their teeth into. Fortunately, the Russell team had delivered up something that fitted the bill exactly.

In the middle of June, William Hardie, the Russell panel's administrator, had emailed his opposite number in Acton's office, Lisa Williams.[335] He explained that he and Russell had been reviewing the minutes of the panel's meetings with UEA staff and that he intended to publish the records of Russell's first visit on 18 January 2010. However, he said he wanted Williams' clearance to publish the notes since they contained

---

*Briffa *et al.* 2008 and Briffa *et al.* 1998.

†That said, since Kelly has made it clear that in the few hours he spent at the unit he saw nothing other than honest attempts to uncover the scientific facts, Oxburgh's observation that the inquiry was unaffected by the Kelly paper may well be correct, something that is perhaps unsurprising given its very narrow remit.

comments about UEA staff and other 'media-sensitive' issues such as the details of Russell's conversations with Norfolk Constabulary.

Williams replied that the verbatim notes had not been intended for publication and suggested that she compose a summary of each interview, pass it to Hardie for review and then circulate it to each of the people involved for final approval. Russell agreed to this idea of rewriting the minutes, adding that the summaries could be 'brief'.[335] Some time afterwards, the minutes were posted to the Russell panel's website.

Hardie's concerns about inadvertently publishing details of the interviews that still needed to be kept confidential were no doubt valid, but it was nevertheless very strange to see an allegedly independent inquiry having to ask for permission to publish details of its work in this way. However, when we note that the minutes of the interviews of CRU scientists and other members of UEA staff all appear to have been taken by Williams rather than by someone working for the inquiry, the alleged independence of Russell's panel does start to look more like a charade than the truth.*[335]

Of course, at the time, these details were unknown to anyone outside UEA and the Russell panel – the emails between Hardie and Williams were only to emerge some months later in response to an FOI request for the university's correspondence relating to the inquiry. However, at the start of August, I received a Google email alert notifying me of the publication of the minutes that they had been discussing. To see a series of apparently 'confidential' documents appear on the web was very surprising and it initially seemed that their publication had been a mistake. However, it soon became clear that most of the contents were of little interest, with one exception: Colam-French's extraordinary revelation that Briffa had taken emails subject to FOI requests home for safekeeping.† Here then was almost incontrovertible evidence that the FOI laws had been breached – at last we had something the headline writers could get their teeth into.

There was considerable debate over whether to change the emphasis of the GWPF report in the light of the Science and Technology Committee's decision to question Oxburgh and Russell about their inquiries – the text was quite critical of the committee's performance, but now that they had shown themselves willing to look at Climategate issues again, there was

---

*Hardie suggested that the notes of the conversation with the police should not be published because the inquiry was ongoing.

†See p. 109.

an understandable wish to 'keep them onside'. In the end, however, it was decided that the text should be left unchanged, and there were only a few last-minute corrections to the text and writing my speech for the press conference to keep me occupied in the days beforehand.

Lawson had arranged for the report to be launched with a press conference in a committee room in the House of Lords, and he would be overseeing proceedings himself. Lord Turnbull, a former head of the civil service, had written the foreword to the report and would speak after Lawson, with me following on behind. Between them Lawson and Turnbull exuded calm efficiency and it was therefore easy for me to be enveloped by a sense that nothing could go wrong. Even the sudden realisation that nobody had sent an embargoed copy of the report to Oxburgh and Russell did little to ruffle feathers and things were soon smoothed over.

When we arrived in the committee room it looked as if attendance would be thin, but there was a last-minute flurry of activity and by the time we got under way it looked as though most of the main UK newspapers would be in attendance. The BBC, however were nowhere to be seen. One of their journalists had telephoned me earlier in the day to see if there was going to be anything new to report and with the 'safekeeping' revelations in mind, I had assured them that the press conference would be newsworthy. Their absence was therefore something of a surprise.

Perhaps the most intriguing attendee was James Randerson, the editor of the *Guardian*'s green webpage, *GuardianEco*, and therefore the man responsible for publishing Ward's hit piece.* Randerson appeared most likely to be the source of difficult questions, although in the event his contributions were limited to asking about how much I had been paid for writing the report and a suggestion that it was hypocritical to accuse Oxburgh and Russell of bias when I had written the report without the input of someone from the other side of the debate. He also argued that I was partisan in the debate – something that was clearly undeniable – neutrals in the climate debate are a rare breed. I therefore agreed with him, but pointed out that this didn't prevent what I said from being true. I could see, however, that Randerson was much animated by my reply and it was clear that he was going to make something of it.

While we were making our presentations, Stringer had slipped un-

---

*We have already come across Randerson as the conduit for the Hockey Team's response to the leak of the emails. See p. 74.

seen into the back of the room, and once the journalists' questioning was complete he introduced himself and explained that he wanted to say a few words about Climategate. He started by defending the Science and Technology Committee, and in particular said that Willis's disparaging description of sceptics as 'deniers' had not influenced the committee's deliberations.* However, having said that he launched a broadside at the state of climate science and the performance of the inquiries. He noted his dismay at the inability of CRU scientists to reproduce their own work and said that it was 'more like literature than science'.† He was also highly critical of the failure of the various inquiries to ensure that none of the issues fell between gaps in their remits, despite the Science and Technology Committee's request that they make sure this did not happen. It was an extraordinary end to the day's events.

The following day, I scoured the newspaper websites to see what they had to say about the press conference and my report. Gratifyingly, most of the coverage was at worst neutral and much of it was supportive. I was surprised to see that none had mentioned Briffa's 'safekeeping' exploits and there was also no mention of Stringer's intervention. However, it was the *Guardian*'s coverage that I most wanted to see, as Randerson was the most likely source of any criticism. In the event, however, although he was quite critical, his article was much less so than might have been expected.[338] He spent much of his article discussing me rather than the contents of the report, but this at least meant that he was not disputing anything I had said. He had to accept that there were serious flaws in the inquiries:

> Prof Phil Jones, the head of UEA's Climatic Research Unit, had been criticised for writing emails about deleting emails that were apparently the subject of a Freedom of Information Act (FoI) request. He has denied deleting any emails to avoid FoI requests, but the main inquiry into the emails conducted by Sir Muir Russell never asked him or any of the other scientists about this.[338]

Coming from such an unexpected quarter, Randerson's belated recognition of the Russell inquiry's failings was welcome. In fact, we had spoken relatively cordially after the press conference, and in particular

---

*See p. 92.

†The only contemporaneous source for this particular remark appears to be my own report on the press conference,[336] although Stringer repeated it in an interview with the *Register*.[337]

about Briffa having taken emails home for 'safekeeping'. Strangely, however, the *Guardian* man had decided not to mention this important new revelation.

Although he had not reported Stringer's part in the press conference, Randerson had managed to get a quote from another MP who had been involved in the Science and Technology Committee inquiry: Evan Harris had lost his parliamentary seat in the election, but was still much in demand from some sections of the media as a talking head on scientific matters. In his comments to Randerson, Harris appeared particularly keen to divert attention from the embarrassing fact of Oxburgh's ignorance about Keenan's fraud allegations:

> Dr Evan Harris, who was a member of the science and technology select committee until he lost his seat as an MP in May, said it would not have been appropriate or practical to address the fraud allegation in detail. 'It would have given weight to [the allegation] that it may not deserve', he said. 'Some of the claims in this area are absurd and, frankly, defamatory. It would have been extremely dubious to have random defamatory allegations dealt with individually'.[338]

It was interesting to see that a fraud allegation that had been taken sufficiently seriously to be reported on the front page of the *Guardian* now being written off as 'absurd' in the pages of the same newspaper. Moreover, the criticism that was being made was that Oxburgh did not even *know* about Keenan's fraud allegation, so Harris's argument that the allegations were baseless did not actually represent any kind of a defence of the inquiry's failure to act. Nevertheless, Harris's intervention was interesting: during the first Science and Technology Committee inquiry he had almost seemed to condone Jones' hiding uncertainties from policy-makers;* now he appeared to be condoning hiding flaws in the science from them as well.

In addition to Randerson's article about the press conference, the *Guardian* had decided to publish a second article concentrating on the report itself. The author, Fred Pearce, was surprisingly supportive, saying that I had raised some valid criticisms of the inquiry. Nevertheless the article made a great deal of what it called my report's 'brazen hypocrisy'.[339] Surprisingly, Pearce emailed shortly afterwards to say that this latter part of the article had not been his work but had been added by the *Guardian*

---

*See quotation on p. 131.

editorial team. It is unclear why Randerson, who was presumably responsible, had chosen to add this to Pearce's article rather than have it appear under his own byline. It was not Randerson's finest moment.

THE RUSSELL HEARING

Coopering up the website

The media coverage of the GWPF report was all out in the open and attention had moved on elsewhere. Although coverage had been broadly supportive, there was no sense of outrage portrayed in any of the articles – there was a feeling that the failures of the Russell and Oxburgh inquiries were entirely to be expected. There were no outraged leader articles, no calls for action from ministers, no demands for a public inquiry. It already looked as if my blog commenter was right – news of the cover-ups was just going to 'fizzle'. Russell's appearance before the Science and Technology Committee was therefore going to be the last chance to change minds about Climategate and the inquiries.

Russell must have been well aware of this fact and he was forced to make some tricky preparations so that his tracks were covered before he appeared in front of the committee. Back in March 2010, after his meeting with the Information Commissioner at UEA, Russell had asked the university for a list of all CRU-related FOI requests received by the university. The list appears to have been provided promptly and it was published on the Russell panel website shortly after the appearance of the report. This site was being carefully monitored for any changes and, soon afterwards was being discussed on sceptic blogs. It was not long before a surprising omission was noticed: the list did not include what was probably the single most important FOI request of them all – request number 08-31, in which Holland had asked for all Briffa's correspondence related to the Fourth Assessment Report and which had led to Jones' now notorious 'delete all emails' request to Mann. As we have seen, Russell had claimed in his report that there was no outstanding FOI request at the time of Jones' email,* and this statement had then been repeated by UEA. So if the full list of FOI requests had been released with the report it would have been hugely embarrassing to both Russell and the university since it would have directly contradicted them.

As soon as McIntyre noticed the critical omission, he wrote to the university to inform them what had happened, asking them to pass on the

---

*See p. 205.

word to Russell. The university responded swiftly and later that same day Williams emailed to tell Russell of what was referred to as 'an important and unfortunate omission'.[340]

Russell was being boxed into a corner. The request's omission was now public knowledge, but if he now posted the missing details on his website the contradiction with what he had written in the report would become obvious. In the event, he appears to have decided that some 'noise' on the blogosphere was a price worth paying and a revised list of FOI requests remained unpublished. However, when the Science and Technology Committee's new hearings were announced, David Holland wrote to the committee to protest at the conduct of the inquiries, and noting in particular the absence of request 08-31 from the published list of FOI requests.[341] With this hot issue now in the hands of the politicians, the university became alarmed and wrote once again to Russell, reiterating that the full list needed to be published. Russell sent an email around his assistants asking if they knew anything about the matter. Finally, a month later and just a few days before Russell's appearance before the Science and Technology Committee, his assistant William Hardie emailed the inquiry's PR consultants asking them to post the revised list on the inquiry website, together with a suggestion for a statement about the implications:

> Please note the addition of the 08-31 FOI request which was previously omitted due to an administrative error. The revision does not affect the conclusion or recommendations of the final report.[340]

Remarkably then, the existence of an FOI request asking for Briffa's IPCC correspondence, dated just two days before Jones' request to Mann to delete all such correspondence (which was entitled, 'FOI & IPCC'), was claimed to have no bearing on the Russell inquiry's conclusion that they had seen 'no evidence of any attempt to delete information in respect of a request already made'.

The hearings expand

The preposterous announcement about request 08-31 was one surprise on the eve of Russell's second appearance before the Science and Technology Committee, but it was not the only one – the parliamentarians had one of their own. Shortly before the new hearing, it was announced that the committee were going to expand the scope of their investigation and that Acton and Davies would be appearing alongside Russell. Although Russell would no longer be answering potentially very awkward questions alone, this did mean that one potential escape route would be cut off: he would no longer be able to say he didn't know the answers or to blame the university.

Deleting emails

When the committee finally reconvened at the end of October, it was clear that deletion of emails was going to be one of the most pressing topics. With the GWPF report published, the news that Russell had failed to investigate this central allegation had been all over the newspapers, and the committee would have been unlikely to miss, for example, the observation by one journalist that:

> Sir Muir seems to have been about the only person studying the affair not to have known about [request 08-31].[339]

The task of questioning Russell about these issues fell to Stringer, who could be relied upon to at least ask some pertinent questions. However, Russell was to prove himself a master of the evasive answer, and he had clearly arrived well prepared.

Stringer opened his questioning by asking about Russell's apparent failure to see evidence of deleting emails subject to FOI; this was clearly contradicted by the existence of Jones' notorious 'delete all emails' request and the recent revelations of Briffa's having taken emails home for 'safekeeping'. Russell replied that Stringer had only quoted half of the sentence in the report, and then unfortunately contradicted himself by quoting what the report actually said – demonstrating in the process that it completely supported the case Stringer was making. However, having regained his composure, Russell went on to raise something he had told the first Science and Technology Committee back in March. During the course of that hearing, Tim Boswell had asked him if he or the ICO was going to investigate possible breaches of the FOI legislation, and he had replied that he was not 'going to put the review into the position of making. . . quasi-judicial prosecutorial, investigative judgments'.[342] At the time, these words had appeared rather vague and had been overlooked, but now, as Russell explained to the new committee what he had meant, his earlier words started to take on a new importance; in fact they started to look as if they had been very carefully chosen:

> ...had we been going to get into this, we would have had to
> start asking questions under caution. We would have been do-
> ing...investigative stuff, because you're getting to the point where
> you're alleging that there might have been an offence, and that
> really wasn't the thing that my inquiry was set up to do... [343]

Russell's remarkable position was that when he had mentioned want-
ing to avoid 'quasi-judicial judgements' he had been telling the committee
that he was not going to investigate the most serious allegations at all: if
conduct might turn out to be criminal in nature, he was not going to get
involved. Instead, he explained, any such allegations should have been
the responsibility of the Information Commissioner.

This was an extraordinary statement, particularly since Russell had
largely given CRU a clean bill of health, reporting that there was no
doubt of their 'rigour and honesty as scientists': this looked like a fairly
empty statement if the most serious allegations had not been addressed.
Stringer certainly appeared astonished by Russell's position and pressed
him to confirm that he had never asked Jones whether emails had been
deleted. Russell's reply left no room for doubt:

> That would have been saying, 'Did you commit a crime?', and we
> would have had to go into a completely different area of the rela-
> tionship and formal role for the inquiry. [343]

Russell was keen to suggest that the matter at issue was not really the
deletion of the email per se, a question he said was not 'actually relevant'.
Instead he suggested, the real issue was 'the end product of the influence
that this process had on what was said in the IPCC report'. It is not entirely clear what he meant by these words. However, he later expanded
somewhat, explaining that the inquiry panel had felt that rather than
examine the email exchanges they should 'look at the question of what
was actually being said in the IPCC report and whether the Wahl and Amman material should be in or not, and what the overall judgment about
that was'. He appeared to be suggesting that since his inquiry had found
that it had been legitimate for Briffa and Wahl to replace the text that
had been considered by the official IPCC reviewers with their own self-
serving text, it was of no consequence whether emails had been deleted
and criminal offences committed by the scientists involved as well. As he
told the committee of his failure to investigate email deletion:

...if we ducked or avoided, I plead guilty to that, but I think we had quite good reasons in terms of our inquiry for not asking that particular question. [343]

Stringer then turned to Acton, asking him if he was satisfied with an inquiry that had failed to ask such fundamental questions. Acton's remarkable response was that the question of email deletion *had* been asked – he had performed the investigation himself:

> It has been investigated. I have asked them and they have assured me that they have never knowingly deleted e-mails subject to a request... [344]

> Can those e-mails be produced? Yes, they can. Did those who might have deleted them say they deleted them? No. They say they did not. I wanted to be absolutely sure of those two, and I have established that to my satisfaction. [344]

Acton went on to explain that a notice to this effect had long ago been placed on the UEA website, an observation that surprised some of those who had been following the story and who knew nothing of this new announcement. However, when the UEA website was examined there was indeed a statement from Jones addressing these issues, just as Acton had said. In it Jones made what appeared to be a clear statement about email deletions:

> I have previously confirmed that I have never knowingly deleted an email that was the subject of an active Freedom of Information request and neither have I deleted data. [345]

However, as McIntyre later observed, it was important to watch the pea under the thimble: Jones was not party to the correspondence most likely to have been deleted: the emails between Briffa and Wahl. So in fact Jones' denial was somewhat beside the point. The important question was what Briffa had done when Jones had asked him to delete his correspondence relating to the Fourth Assessment Report.

Stringer pressed the point, asking Acton if he had asked Briffa why he had taken emails home, but the UEA man appeared to be prepared.

> I didn't. I can, if it is appropriate, tell you an element that I think may bear upon it, which was that at the time he was gravely ill and rather frequently not in the university. So to take a copy home does not seem to me very extraordinary, but I haven't asked him. [344]

Acton's story is somewhat implausible – the word 'safekeeping' is quite incongruous with his explanation – but he had no basis for this suggestion anyway; he was quite clear that he was speculating. However, he explained that all of the emails referred to were still in existence, thereby implying that nothing untoward could have happened. He was supported in this by Russell who confirmed that the emails in question formed 'part of the 3000 pages',[344] in other words the Climategate disclosures.

These were surprising statements, at least to McIntyre. Six months earlier, he had sent yet another FOI request to UEA asking for attachments to those Climategate emails relating to the Wahl and Ammann paper. These, he hoped, would reveal important details of the IPCC review of the Hockey Stick papers. However, the university had rejected McIntyre's request on the grounds that the information was 'not held'. The contradiction between the university's response to McIntyre and what Acton was telling the Science and Technology Committee was stark and McIntyre responded by appealing to the Information Commissioner.

At time of writing this appeal has yet to be heard. However, information that has come to light since Acton gave his evidence to the committee has provided an explanation for the discrepancy, as well as putting Acton's words in a disturbing light. As we now know, when Holland requested Briffa's IPCC correspondence, CRU staff responded by moving the relevant emails to memory sticks. Everyone involved in the Science and Technology committee hearing had at least some awareness of what had happened, since the 'safekeeping' revelations were now public knowledge, although the record of the meeting at which Colam-French gave the Russell inquiry this important information did not specify precisely which emails were involved. This is an important question, because Acton's suggestion that Briffa had been unwell and simply wanted his emails to be available at home was not entirely borne out by the known facts. Although it appears clear that emails were moved *en masse* to Briffa at the time of his illness,[346,347] this was in the second half of 2009, more than a year after he had concealed his IPCC correspondence. So while Acton's story was true, it was grossly misleading, since Briffa's indisposition could not have explained the concealment of the IPCC emails the year before.

The explanation for the contradiction between UEA telling McIntyre that the email attachments were 'not held' and telling Parliament that everything was still available lies in an obscure detail of FOI legislation. Since the FOI Act and the Environmental Information Regulations came into force, the courts have established that information held on backups is 'not held' for the purposes of the FOI Act – in other words, if information has been deleted from normal storage locations, public bodies are not required to search backup and disaster recovery systems for it. So if the CRU scientists had moved the information that McIntyre had requested to a memory stick, or simply just deleted them, UEA might have felt justified in rejecting his request because the information was only 'held' on backup servers – although, of course, they would not have told McIntyre that this was their reasoning. Acton could, at the same time, have told the committee that the emails were all available.*

However, all these details of McIntyre's request and the underlying legal questions were, of course, unknown to the committee, who were no match for the evasions of the wily bureaucrat. Acton's story, like Russell's before it, went completely unchallenged.

The Oxburgh remit

Although the second hearing was expected to concentrate on the Russell inquiry, the addition of Acton and Davies to the list of witnesses gave the committee the opportunity to follow up on some of the questions raised by Oxburgh's appearance before them a few weeks earlier. Committee chairman Andrew Miller launched straight in by repeating the error that had been made by the BBC's Roger Harrabin, asking Acton whether the terms of reference for the Oxburgh inquiry had been changed rather than whether they had been misrepresented to Parliament. Acton's reply was straight out of the *Yes, Minister* school of obfuscation.

> From my point of view, no, it wasn't. What I think I said to your predecessor committee was that the purpose was to reassess the science and see if there was anything wrong. Having observed that this question might be raised, I wanted to check how your predecessor committee had understood me. I was pleased to see that in the note, paragraph 131, your predecessor committee says that what this panel should do is determine whether the work of CRU has been soundly built. That is exactly what I meant. Was it scientifically justified? [348]

---

*There was, however, a flaw in this logic: McIntyre's request was submitted under the Environmental Information Regulations rather than the FOI Act, and so the legal precedent about information only held on backup systems did not necessarily apply. In fact, the Information Tribunal has subsequently ruled on a related case that UEA is required to search backups.

Try as they might to pin him down, Acton was far too slippery for the members of the committee and as he threatened to launch into a discussion of what he called the 'protean' word, 'science', it was almost possible to sense eyeballs rolling around the table.

Oxburgh's eleven

Perhaps wisely, Miller decided to hand over to Stephen Metcalfe, who moved the questioning on to the related subject of the list of papers selected for Oxburgh's team to examine. Acton freely admitted that the list was essentially the same as the list that had been included in the UEA submission to the first parliamentary inquiry, claiming that it would be 'very odd to draw Lord Oxburgh's eyes away from that list when they seemed to be bang on the issues at stake'. This was confirmed by Davies, who explained his thinking in putting that list together:

> They were chosen to address... the huge unfounded criticism of a number of areas: the CRU global land temperature records, so there were two publications from that area; homogeneity adjustments – two publications from that area; urbanisation effects – two publications from that area; tree ring density records – three publications from that area; and then accusations of cherry-picking long records of tree growth, and there were two articles from that area. [349]

Stringer later pressed Davies on the same point, noting that the papers that were at the centre of the controversy – the multiproxy temperature reconstructions – had not been included on the list.

Davies' response was astonishing and utterly shameless:

> I would dispute that. [350]

He then launched into an explanation so obscure that there can be little doubt that the smokescreen was a deliberate tactic. Even with the benefit of a transcript of his words, the meaning is sometimes hard to discern:

> One of the few comments [about which papers should have been included] has been in Mr Montford's report and he mentioned three publications in particular: Jones *et al.* of 1998, Mann and Jones, 2003, and Osborn and Briffa 2000.* That list seems to have been taken from a longer list on Mr McIntyre's blog site of 15 April

---

*Sic. Osborn and Briffa was published in 2006.

after the Oxburgh Report came out. In fact, the majority of the publications listed by Mr McIntyre on 15 April were referenced in the CRU's submission to Muir Russell, which the Oxburgh Panel received, or some were on the Oxburgh starting list.[350]

McIntyre had indeed listed those papers about which he was concerned in a blog post shortly after the Oxburgh panel had reported. The papers he discussed are set out in Table 11.3, which shows that there was almost no commonality with the studies on the list sent to Oxburgh. Even a cursory glance shows that Davies had no basis for disputing Stringer's observation that none of the multiproxy papers were on the Oxburgh list. In fact, four of the five tree-ring papers on Oxburgh's list had been discussed in favourable terms in a separate article McIntyre had written in May 2005. As McIntyre noted, it was rather extraordinary that almost none of the papers he had criticised had been included on Oxburgh's list but all of the ones he had raised no concerns about had.

Davies appeared to be trying to make a case that since the papers McIntyre was concerned about were included on CRU's submission to the Russell inquiry, their absence from the list sent to *Oxburgh* was excusable. He also seemed to be claiming that the Oxburgh panel had 'received' a copy of the list sent to Russell. There were, however, a number of problems with this argument. Firstly, there were almost 120 separate citations made in CRU's submission to Russell, 70 or more of them being CRU papers. Assuming the Oxburgh panel members had in fact made use of the document, they would have found it next to useless in terms of guiding them to the pertinent issues, given how little time they were given for their inquiry. But what is worse, as we have seen, the email correspondence of one Oxburgh panel member – David Hand – was obtained under FOI and there is absolutely no sign of him having received CRU's Russell submission. The Oxburgh report made no mention of having received the document either.

If this masterly display of obfuscation were not enough, Davies had a whole new layer of confusion to add. He raised the subject of McIntyre's submission to the first Science and Technology Committee inquiry, and claimed some sort of vindication in the fact that there was some commonality between the papers McIntyre had mentioned there and the papers on the Oxburgh list. The centrality of the multiproxy studies to McIntyre's work was not mentioned. Even then Davies was not done, listing some of the papers in McIntyre's submission and trying to persuade the

TABLE 11.3: Papers examined by McIntyre, Oxburgh and Russell

Paleoclimate papers		Considered by		
		McIntyre	Oxburgh	Russell
Briffa et al. 1992	Tornetrask, Briffa bodge	×		×
Briffa et al. 1995	Polar Urals	×		×
Briffa 2000	Yamal, Taimyr	×	×	×
Briffa et al. 2002	The cargo cult 'assumption'	×		×
Mann and Jones 2003	Multiproxy	×		×
Jones and Mann 2004	Multiproxy	×		
Osborn and Briffa 2006	Multiproxy	×		×
Rutherford et al. 2005	Multiproxy	×		×
Jones et al. 1998	Multiproxy	×		×
Briffa et al. Nature 1998			×	×
Briffa et al. Phil Trans B 1998			×	×
Briffa et al. 2001			×	×
Briffa et al. 2008			×	×

Surface temperature papers		Considered by		
		McIntyre	Oxburgh	Russell
Jones et al. 1990	The urban heat island paper	×	×	×
Jones et al. 2008			×	
Jones and Moberg 2003			×	
Jones et al. 1986a			×	
Jones et al. 1986b			×	
Brohan et al. 2006			×	

The table shows the papers identified by McIntyre as being the ones most criticised by him, which are all tree ring papers apart from Jones et al. 1990. Also shown are those papers that were also on the Oxburgh list and those that were also on the CRU submission to Russell.

Committee that the issues had been covered by Oxburgh. For example, he claimed:

> Mr McIntyre also mentioned Briffa *et al.* 1992. That was not on the starting list. This was a paper on the Torneträsk tree-ring series. This was covered in Briffa 2000, which was on the starting point list, and it was referred to in the Muir Russell submission. The only other paper which McIntyre mentioned was Briffa *et al.* 1995. That also was not on the starting point list. This was a paper on Polar Urals tree rings. This was covered in Briffa 2000, which was on the starting list, and was also referred to in the Muir Russell submission.[343]

The two earlier papers – Briffa *et al.* 1992 and Briffa 2000 – were paleoclimate papers, in which McIntyre had discovered some serious issues: a notorious ad-hoc adjustment known as 'the Briffa bodge' in one and the suppression of new data that didn't support the global warming hypothesis in the other. Both were component series in the Briffa *et al.* 2000 temperature reconstruction, but they were taken as finished entities for that paper, so the bodging and the suppression of data were therefore invisible to its readers. So when Davies told the committee that the two papers were 'covered', he was leading them to believe that that the concerns McIntyre had raised would have been considered by Oxburgh and his team. This could not have been further from the truth.

The Royal Society

The question of who had chosen the papers for the Oxburgh panel and precisely what role the Royal Society had taken in approving them was also one that attracted the attention of the committee members. In answer to a question from Stephen Metcalfe, Davies explained that he had in fact coordinated the selection of the papers himself, and he volunteered the new information that he and the acting director of CRU, Professor Peter Liss, had met with Lord Rees approximately a week before the email exchanges in which Davies sought the use of the Royal Society's name. As Davies told it, he had sent the list to the Royal Society on 4 March 2010, a week before it was sent to the panel members. There are hints in the email correspondence that this may be correct. Certainly in the email where Davies first broached the subject of using the Royal Society's name, he said 'I did send you this list earlier. . .',* although it is

---

*See p. 151.

surprising that the actual email – assuming it was sent that way – has not been released in one of the many FOI requests that UEA has responded to since Climategate. Whatever the truth of the matter, there is still the unavoidable fact that the list of papers had already been sent out to the panel members before Brian Hoskins had approved it on behalf of the Royal Society.

Metcalfe continued to push for details of how Oxburgh's list was put together, asking whether Jones had been involved in the selection. Davies' response was gratifyingly brief, but intriguing.

> No, not for the Oxburgh Panel. The discussions were internally at UEA between me and Professor Liss. Professor Jones and his colleagues were told which publications would be sent in and would be recommended to the Oxburgh Panel, but they had no decision-making role at all. [349]

Davies' response appears to be carefully worded. He had already noted that the Oxburgh list was in essence the same as the CRU submission to the first Science and Technology Committee inquiry. Jones may well have had no involvement in the preparations for the Oxburgh inquiry, but by carefully restricting his response in the way he did, Davies may well have been drawing a veil over Jones' involvement in selecting the papers sent to Parliament six months earlier. Certainly, given that much of that document was billed as being CRU's response rather than being something written on behalf of the university as a whole,* it is hard to imagine that Jones did not prepare it himself. The conclusion seems clear: the papers that were ultimately examined by the Oxburgh inquiry were probably selected by Jones himself.

Peer review

At the time of the hearings, Pamela Nash had been a member of Parliament for just five months and was, at 26 years old, the youngest member of the House of Commons. Her focus was on the knotty issue of peer review, and she opened her questioning by inquiring why the three particular instances of alleged peer review abuse examined by Russell and his team had been chosen. Russell's response was rather strange:

> They were the three that had been at the top of the head, as it were, in the comments that were made when the whole story broke. I

*See p. 127.

keep going back to what I said to Mr Williams. They were the things which we thought, as we were looking at the issues, were solid and good examples to pick and to test the accusations that had been made. [351]

It is not entirely clear what Russell meant by 'at the top of the head'. However, in rounding up his response on peer review, Russell used this strange form of words once again:

I put my hand up and say, yes, there could well have been other cases that we might have looked at, but these were the ones that everybody seemed to think were at the top of their heads at the time. [351]

In context, Russell's choice of words appears to be a careful one, allowing him to insinuate that he had selected the allegations that were most widely discussed at the time, without actually saying so clearly. The episodes he had chosen for investigation* had certainly attracted little or no attention beforehand, with only the Cook review having been the subject of any commentary at *Climate Audit*. The insinuation was therefore not true, but the committee members could not be expected to see through Russell's misinformation.

Stephen Mosley took up the reins on the peer review issue, asking in particular about the possibility that Jones had breached peer review confidentiality. As we have seen, the Russell panel's own adviser had suggested that this was an area of concern.[†]

We didn't dig into that one at the time. I acknowledge that we didn't. I challenge the proposition that we were recommended to do it by anything that was said in the Horton paper. There is a statement from Mr Montford that we ignored the advice of our expert and didn't follow up on this, but I don't think that is actually right. So I will readily admit that that was not one that we chased down. [352]

As we have seen, Horton's paper was about the generalities of peer review rather than the specifics of what happened at CRU. But given that Horton had raised this area as an ethical concern, it is still remarkable that the inquiry didn't follow it up. That Russell and his team were

---

*See p. 180 *et seq.*
[†]See p. 182.

happy to use Horton's observations about the controversial nature of peer review, but ignored his observations about confidentiality looks opportunistic, to say the least. However, Russell had at least admitted that another important area of concern had not been addressed.

Then, perhaps trying to limit the damage, Davies interjected that it is quite normal for scientists to seek outside help on particular aspects of a paper they are reviewing. This may or may not be true, but it is hard to equate this story with the Jones correspondence in question, in which he stated that he wanted to make sure that his views were incorporated in *Osborn's* review of a paper.

Keenan and the China stations

With Oxburgh having embarrassingly admitted that he had been completely unaware of the fraud allegation that Keenan had made against Jones, the committee were keen to know if Russell had looked at the important question of what Jones had known and said about weather station moves in China, and when. Metcalfe asked Russell if the inquiry had questioned Jones directly about the allegation, but the answer must have left him very confused:

> Yes. As far as I recall, we did. I am not just sure whether we picked that one up in one of the interviews or not, but we had plenty of information about it. [352]

So when he answered 'Yes', his subsequent statements in fact suggested that he was merely saying 'maybe' – presumably he was not sure because he had failed to attend the interviews with Jones himself. Later in the session, however, Davies went even further and stated that the allegations of fraud had been 'investigated thoroughly'. [353] In fact, the only time Jones appears to have been asked about his 1990 UHI paper was during Norton and Clarke's visit at the start of March,* when he admitted that the China stations *had* moved, although he 'did not believe... that there had been gross moves of location'. [233] Russell and Davies appeared to believe that since Jones' later paper on UHI reached the same conclusions as the controversial 1990 paper, there was no case to answer. That Jones had been citing the 1990 paper, in the full knowledge that it was flawed, before his alleged confirmation was published went un-

---

*See p. 167.

mentioned. Likewise, the fact that the 'confirmation' relied on the same flawed set of China stations was sidestepped.

Davies followed up with his own observations on the fraud allegation. Noting that Jones' co-author Wang had been cleared of wrongdoing by his university, he went on to say:

> The two publications in question were subject to assessment by Lord Oxburgh and they saw nothing wrong at all. It is perfectly true, although Mr Keenan did not do any data analysis whatsoever in this article in the journal, that he did make some claims about the stations having moved location. Phil Jones acknowledged that the 1990 paper could be improved as a result of this further knowledge. He had no access to the knowledge at the time and, as Sir Muir has indicated, did in fact confirm the results of the 1990 paper in a later paper in 2008.[353]

The 'two publications' were presumably the Jones 1990 UHI paper and Wang paper on which he had been co-author. However, the idea that the two publications were 'subject to assessment by Lord Oxburgh' who had seen 'nothing wrong at all' is hard to comprehend. Just a few weeks earlier, Oxburgh had made it quite clear to the committee that he knew nothing of the Keenan allegations and it is therefore highly unlikely that he knew about the Wang paper either. This was all extraordinarily misleading, but once again, the wall of obfuscation appeared to do its job and the committee moved on to other matters.

THE CLIMATEGATE ANNIVERSARY

When the second hearing closed, the committee went quiet and disappeared, leaving all the onlookers to wonder what they were up to. No hint emerged of the discussions they were having and in fact the committee's minutes reveal that there was little consideration of the Climategate inquiries during November 2010. In the meantime the anniversary of the disclosure of the emails passed by with a series of articles looking back over the previous twelve months.

Perhaps most notably, two media outlets that had been firm upholders of the global warming orthodoxy decided that perhaps a more balanced line was called for, at least this once. Harrabin, speaking to a conference at an environmental conference said:

> [Then there were] the shenanigans around Climategate, into which I have looked very closely, and I can find no smoking gun, but the

investigations into Climategate have really been rather inadequate, I think, from the point of view of the public's expectations of what they would produce, and there has been a loss of public confidence in climate science...[354]

The *Economist*, meanwhile discussed my report in terms that at least seemed to recognise the validity of what I had been saying:

> Andrew Montford, a blogger with the same predispositions as the Foundation, sums up the principal Climategate allegations in a way that shows them to be much more about process than about manipulated findings. He cites an exclusion of sceptical views from the literature; a misrepresentation of primary research, and its uncertainties, in some secondary presentations; a lack of openness to requests for information and a willingness to contravene Britain's freedom of information act; a discordance between what the scientists said in private and what they said in public. Fraud in basic science and primary data ... which is commonly said to have been revealed, does not make the list.[355]

This was fair comment, although if indeed it was accepted that sceptical views had been excluded from the literature and that primary research was misrepresented in the secondary presentations, the implications on the validity of both the primary literature and the IPCC reports for policy-making purposes are rather profound, and were surely worth a mention in the article.

THE PRESSURE

In December a series of letters was sent to the committee by some of those who had complained about the conduct of the inquiries. The first was written by me,[356] outlining in bullet point format the areas that were undisputed, those that were disputed – including what I hoped were decisive reasons for the committee to accept my take on events – and lastly those areas that did not even appear to have been investigated. The same day, Keenan[357] wrote setting out his fraud allegation and asking for the committee to set up an inquiry. The result of these efforts is not recorded, beyond the fact that the committee asked for them to be published alongside the other evidence.

It was not until the New Year that Miller and the other committee members brought matters to a close with a meeting to discuss the draft report. The minutes record that only five of the eleven members of the committee felt it necessary to attend – Miller, Stringer, McClymont, Mosley and Metcalfe.

We know little of what was said – the deliberations of parliamentary committees are a closely guarded secret and the official minutes record only the barest details of what takes place. However, it is possible to discern that Stringer tried to persuade the committee to consider the Kelly paper, and all its damning criticisms of the way science was conducted at CRU. With it being some four months since Kelly's paper had first been raised at Oxburgh's hearing, it is extraordinary that this vital piece of evidence was only now being considered. However, as we shall see, they appear to have been unmoved by it anyway.

Stringer may well have known by then that the other committee members were resolved to let CRU off the hook, but he was at least going to have his objections set down in writing, and he put forward an amendment to the report, which is a powerful summation of the issues:

> There are proposals to increase worldwide taxation by up to a trillion dollars on the basis of climate science predictions. This is an area where strong and opposing views are held. The release of the e-mails from CRU at the University of East Anglia and the accusations that followed demanded independent and objective scrutiny by independent panels. This has not happened. The composition of the two panels has been criticised for having members who were over identified with the views of CRU. Lord Oxburgh as President of the Carbon Capture and Storage Association and Chairman of Falck Renewable appeared to have a conflict of interest. Lord Oxburgh himself was aware that this might lead to criticism. Similarly Professor Boulton as an ex colleague of CRU seemed wholly inappropriate to be a member of the Russell panel. No reputable scientist who was critical of CRU's work was on the panel, and prominent and distinguished critics were not interviewed. The Oxburgh panel did not do as our predecessor committee had been promised, investigate the science, but only looked at the integrity of the researchers. With the exception of Professor Kelly's notes other notes taken by members of the panel have not been published. This leaves a question mark against whether CRU science is reliable. The Oxburgh panel also did not look at CRU's controversial work on the IPCC which is what has attracted most series [sic] allegations.* Russell did not investigate the deletion of e-mails. We are now left after three investigations without a clear understanding of whether or not the CRU science is compromised. [358]

---

*This should presumably read 'serious'.

The minutes record that McClymont, Mosley and Metcalfe decided to vote against the amendment, with Stringer's the sole vote in favour.

The report appeared a week later, with a distinct lack of fanfare; it was almost as if the committee wanted to put Climategate behind them as fast as they possibly could. A few scraps were thrown the way of CRU's critics, but overall it was very much a case of 'move along, nothing to see here'.

On the positive side, there was a recognition that UEA had been misleading over the nature of the Oxburgh inquiry. The committee were also critical of the way Oxburgh had conducted his inquiry, both in terms of its lack of independence and the way it completed its work in a matter of days – as they perceptively noted, UEA's demand that the inquiry be completed quickly strongly suggested that Oxburgh's work was anything but independent.

However, elsewhere it was almost as if the evidence the committee had heard had washed over them without registering any reaction. Oxburgh's new claims that the criticisms in the Kelly paper were about 'climate science as a whole' rather than about CRU in particular had found their way from the footnote in the transcript to a paragraph in the main text of the report. The claim was reproduced verbatim and without comment by the committee. By failing to question the truth of Oxburgh's statement – something that could have simply been done by examining the Kelly paper itself – the committee implied to the reader that Oxburgh was correct. They then reinforced this position by insinuating that the other panel members' working papers would have presented a different view of CRU's science. These, they said should be published forthwith, although the existence of any other such working papers appears to be largely speculation, with Oxburgh's evidence silent on the subject and with no such papers having appeared in the period since the committee's request.* Then, as if attempting to finally kill off any further discussion of Kelly's paper, the committee made a remarkable assessment of what it meant:

> The importance of Professor Kelly's work is that it clears CRU of deliberately falsifying their figures but, as the [Oxburgh] report

---

*In fact, at least one panel member – Kerry Emanuel – destroyed his working papers soon after the completion of the inquiry.[359]

put it, 'the potential for misleading results arising from selection bias is very great in this area'.[360]

This spin on the Kelly paper appears to veer between highly debatable and outright false. It is surely likely that very little reassurance can be taken from Kelly's observation that he had seen no sign of wrongdoing, given the few hours the Oxburgh team actually spent at UEA. As for the committee's claim that Kelly's concerns had been taken on board in the report, this needs to be compared to what Kelly actually said, namely that the potential for 'conscious or unconscious bias in the choices of data' could not be eliminated. In other words *conscious* bias – cherrypicking – had not, in Kelly's view, been ruled out. It is therefore remarkable that Oxburgh's team did no further work in this area, and more remarkable still are the committee's conclusions that Kelly's concerns had been covered in the report.

Elsewhere, the failure of Russell and his team to perform even a cursory investigation of the most important allegations were given the seal of approval of the committee: despite failing to ask what Jones had said to journal editors during the Soon and Baliunas and Saiers affairs, the committee reported themselves 'satisfied with the detailed analysis of the allegations'. Of the failure of the inquiries to ask about deletion of emails they stated that they were 'concerned', although they clearly derived considerable comfort from Acton's statement that Jones had told him that he had deleted nothing. The committee, it must be said, were probably on shaky ground in this area, having themselves failed to ask this question of Jones when they had the opportunity.

If the panel were 'concerned' by the failings in the area of FOI, when it came to the integrity of the Oxburgh panel they were 'frustrated', noting the failure to look at the multiproxy papers that had been most criticised, but claiming that the papers that Oxburgh had chosen to look at 'were central to CRU's work and went to the heart of the criticisms directed at CRU'. This was clearly not true – as we have seen the papers chosen were almost all those that had not been criticised at all. In fact, the committee seemed to want to suggest that because some paleoclimate papers had been examined, the failure to look at the critical multiproxy papers was somehow forgiveable.

Vague statements of concern might have been enough on these topics, but on the allegations concerning Jones' UHI paper there was no alternative but to make a deft sidestep and the committee accepted Russell's arguments that Jones' 2008 paper was somehow capable of exonerating him of having cited his flawed 1990 paper in the 2007 Fourth Assessment Report. Once again it is very hard to come away with a favourable impression of the committee's actions.

LOOSE ENDS

The story of Climategate is still unfolding and it remains to be seen whether the last has been heard of the Climatic Research Unit and the inquiries that tried to cover up their wrongdoing. Here I will briefly mention some recent developments that may turn out to be important.

At the end of 2011, RC made a second appearance, posting a new and much larger dossier of emails from the CRU onto the internet. The media reaction was muted, but in fact there were some important revelations, several of which have informed this book. Among these were the fact of Jones' meeting with von Storch ahead of the resignation of the board of *Climate Research*, Briffa's rejection of the Hunzicker and Camill paper, but perhaps most importantly, the clear evidence that Jones and his colleagues had been hiding emails on memory sticks rather than deleting them.

Despite the lack of interest from the mainstream media, the new disclosures did not go unnoticed. The House of Commons Science and Technology Committee are understood to have held a third inquiry into Climategate, inviting Acton to appear before them again for a short hearing. However, this time the evidence was all taken in camera and no record of what was said has appeared to date.

Meanwhile, investigations into the activities of CRU and the inquiries have continued unabated. McIntyre was able to obtain details of Briffa's Urals chronology through an FOI request, uncovering in the process all of the deceptive statements that Briffa had made in public and to the Russell inquiry. David Holland's work has also continued, but his efforts have been resisted at every turn. An attempt to obtain Russell's email correspondence under FOI has been refused and the appeal process is ongoing. Meanwhile, Edinburgh University, which employs both Geoffrey Boulton and Jim Norton, has reported that all correspondence related to the Russell inquiry was deleted from their servers within days of the publication of the report.

Holland has also pursued the UEA's false representation to the Information Commissioner that he had appealed their decision on Overpeck's spreadsheet.* In the Spring of 2012 he issued a formal complaint to Norfolk Constabulary alleging that UEA's falsehood amounted to a perversion of the course of justice. While preliminary inquiries were made, no investigation was launched, apparently because the ICO could not 'evidence any intention to commit offences by any party'.[361] In other words, there was no evidence that the falsehood was intended to avoid further investigation of FOI offences, although what other purpose it could have had is hard to discern.

Across the Atlantic, parallel efforts to obtain Michael Mann's emails have met with mixed success. The first, by the Virginia Attorney General, Ken Cuccinelli, was struck down by the Virginia Supreme Court, which ruled that Cuccinelli had not demonstrated any grounds for his investigation. A second attempt, by an advocacy group called the American Tradition Institute, is ongoing and may well stand a better chance of success.

The corrosive effect of Climategate on the reputation of climatology and of science in general continues unabated.

---

*See p. 200.

CHAPTER 12

MENDACITY, FAILURE, AND THE PUBLIC INTEREST

FAILURE OF THE INQUIRIES

The Science and Technology Committee, the Russell inquiry, Oxburgh's inquiry, the Penn State inquiry, the Penn State investigation: every official panel that has been asked to assess the Climategate emails has issued a 'not guilty' verdict – either a full one or one that was only slightly leavened with criticism of the CRU scientists. The headlines that have greeted each of the reports have all been of names cleared and reputations restored, of critics sent packing and of science reaffirmed.

Yet as reports of each successive whitewash have hit the news stands there has been little sense that anyone has been convinced. Members of the public are not fools. If a public institution launches an investigation of its own staff's conduct the results will be heavily discounted, even at the best of times. If on the other hand the investigation is internal, operates under absurdly restrictive terms of reference, fails to interview critics and fails to examine the allegations in a meaningful way, the results will be tossed aside as worthless, even if many environmental journalists are willing to argue otherwise in the service of the greater green cause.

The failings of the inquiries are transparent and simple to understand. For example, the refusal of Russell and his colleagues to perform a meaningful investigation of email deletions is stark and unarguable – it has been left to bloggers to uncover the truth of the matter and a surly silence has been the only response from the majority of the mainstream media. Likewise, despite the perversion of the peer review process having been possibly the most important allegation to emerge from Climategate, we still do not know if the Hockey Team followed through on their plans to force *Climate Research* and GRL to come to heel. Not one of the inquiries saw fit to ask this most simple question: 'Did you contact these journals and, if so, what did you say?' Instead the Russell panel, the Penn State investigation and the Science and Technology Committee accepted the word of those accused that they had done nothing wrong. The majority of the mainstream media are damned by their refusal to report these failings and by their repeated assertions that the CRU scientists have been cleared.

Although there are many in the media who would like to draw a veil over Climategate and the cover-ups that followed, word has now begun to spread that all is not well with the science that underpins carbon policy. Fred Pearce, who we met in Chapter 11, is probably without peer among environmental journalists and, as we have seen, has been quite clear that the inquiries failed to get to the bottom of the allegations made against CRU. Even the BBC's Roger Harrabin has gone on the record along similar lines:

> Climategate was a real problem for the public consciousness. It seemed like something dodgy had gone on. Now I've looked very deeply into Climategate and I can't find any smoking gun at all. But I've also followed the enquiries into Climategate, and in my view they were all inadequate. [362]

The failure to get to grips with misconduct by global warming scientists has had other important implications, not least of which has been continued concern over the integrity of the peer review process. A case in point was the publication of a paper by Roy Spencer and William Braswell in the journal *Remote Sensing* in the autumn of 2011, the latest instalment in a long-running battle between these two sceptics and a climatologist of more mainstream views named Andrew Dessler. The Spencer and Braswell paper was the subject of huge media interest, since it appeared to undermine the global climate models and suggested that projections of global warming were overstated. Soon afterwards, however, there was a dramatic development when the journal's editor, Wolfgang Wagner, declared that the publication of the paper had been an error. He said that he took full responsibility for the decision to publish it and announced that he was resigning:

> Peer-reviewed journals are a pillar of modern science. Their aim is to achieve highest scientific standards by carrying out a rigorous peer review that is, as a minimum requirement, supposed to be able to identify fundamental methodological errors or false claims. Unfortunately, as many climate researchers and engaged observers of the climate change debate pointed out in various internet discussion fora, the paper by Spencer and Braswell that was recently published in *Remote Sensing* is most likely problematic in both aspects and should therefore not have been published.
>
> After having become aware of the situation, and studying the various pro and contra arguments, I agree with the critics of the paper.

Therefore, I would like to take the responsibility for this editorial decision and, as a result, step down as Editor-in-Chief of the journal *Remote Sensing*.[363]

It appears then that Wagner's decision to resign was, somewhat incredibly, based on things that he had read on 'various internet discussion fora'. However, shortly afterwards a more sinister explanation was suggested when it was revealed that Wagner and the publisher of *Remote Sensing* had both written personal letters of apology to Kevin Trenberth. This was extraordinary because Trenberth was not actually a party to the dispute between Spencer and Dessler, although he is, as we have seen, a leading figure in the Hockey Team and an influential figure in the IPCC.

It appears then that the major achievement of the inquiries run by Penn State, UEA and the Science and Technology Committee has been to allow free rein to this kind of distortion of the normal workings of the scientific process.

This is not, however, to suggest that this was their motivation in covering up the misdeeds of the Climategate scientists. There are almost certainly more mundane incentives at work in the minds of those who commissioned and led the inquiries.

THE ECONOMICS OF THE ACADEMY

There can be little doubt that both UEA and Penn State had strong incentives to cover up the misdeeds of their staff. Jones and Mann were star performers in their field, bringing a wealth of friendly media attention to their universities. The attention lavished on star scientists can be of enormous benefit to their employers, nudging them into a virtuous circle of attention and funding: the scientists win citations in the literature, which in turn promotes their employers in assessments of research excellence, bringing in additional funding, and so on *ad infinitum*. Jones alone had brought in an average of £700,000 in grant funding each year for the 20 years prior to Climategate,[364] so there is no doubt that the sums involved are quite large. Moreover, if there is any doubt of the strong incentives to cover up the misdeeds of star performers, a recent incident at Penn State lays these to rest. In November 2011 it was announced that Penn State President Graham Spanier, who we met in Chapter 6,* had been ousted as the result of an allegation that one of the university's football coaches had been abusing children in his care. Spanier was accused of making

*See p.123.

an inadequate response when the abuse was brought to his attention, fearful of a loss of funding to the university's football programme.

While the allegations against climatologists are trivial by comparison to those swirling round Penn State's football programme, the lessons seem clear: the incentive structure for a state-funded university is such that senior officials may be capable of covering up even the most serious crimes. The cost of finding miscreants guilty is high – reputations damaged, funding lost and so on – and the risk of discovery of a cover-up is small. Without a whistleblower, it is highly unlikely that the truth would ever see the light of day.

OVERSIGHT

If universities will protect their own interests ahead of those of the public that pays their salaries, who then will ensure good behaviour among scientists? The evidence of Chapters 7 and 11 is that we should not look to elected representatives for help, since the Science and Technology Committee inquiries were at best inadequate and at worst a pair of whitewashes.

As we look back at the performance of the first Science and Technology Committee, many aspects of the way their investigation was conducted appear disturbing. Why were McIntyre and McKitrick not invited to give oral evidence? Why was so much time spent on the peripheral issue of CRUTEM? Why was Jones cleared of the peer-review allegations based only on his own denial of wrongdoing? Why was Jones not questioned about email deletions? Why was the Saiers affair not examined? How did the committee manage to miss McKitrick's allegation of fabrication? The list goes on and on.

The committee might have some defence in the fact that their time was limited by the approach of the general election, but it is likely that some of the shortness of time was self-imposed: the report was published two full weeks ahead of the dissolution of Parliament, so it would actually have been quite possible for the panel to perform their investigation much more diligently than they did in practice. The fact that only a single morning of hearings was held looks highly problematic. To put it in context, on the committee's comparatively trivial inquiry into peer review during 2011, they heard from as many as 23 witnesses over four days.

Even if time really was so short that a thorough investigation was impossible, it would still have been possible for the committee to issue an interim report and to ask the successor committee to pick up the inquiry again after the election. Instead a decision was taken to issue a full exoneration, despite the fact that no meaningful investigation had taken place. In these circumstances the cursory nature of the committee's look at the Climategate allegations looks opportunistic, to say the least.

The second Science and Technology Committee inquiry looks little better. Here there was no time pressure and the issues were much more simple. But once again there was no opportunity for the principal critics of Russell and Oxburgh's work to make their case – there had been no opportunity for anyone to put the 'case for the prosecution' at all. Instead we were presented with the unedifying sight of the politicians being over-whelmed by a tsunami of distortions, misrepresentations and outright falsehoods from those behind the investigations. The committee mem-bers, several of them new to Parliament, were singularly ill-equipped to deal with this kind of behaviour, although, like their predecessor com-mittee, they made a rod for their own backs by failing to hear from CRU's critics.

Despite the failings of the second Science and Technology Committee inquiry, we might still be prepared to accept that we were dealing with the results of incompetence rather than malfeasance, were it not for the committee's treatment of Stringer's amendment.* As we have seen, this was a carefully worded proposal that asked the committee to conclude only the patently obvious – that the inquiries had been flawed and that important questions had not been answered. After all, if sceptics and green journalists such as Roger Harrabin and Fred Pearce could agree on this position then it is hard to see it as a particularly controversial one. But despite the evidence that they had heard, the members of the committee chose to vote the Stringer amendment down, something that suggests strongly that there was an agenda other than truth-seeking at work.

We cannot, of course, know what motivated the committee members to take the course they did, but we can at least speculate. It is possible that the motivation was simply grubbing after votes – and in particular the votes of the environmentally minded. The green electorate has been hotly fought over for several years, and in the run up to the election, politicians will have needed little encouragement to put this highly vo-cal group first. But even after the election, with a Conservative/Liberal

*See p. 262.

Democrat coalition in power, there may well have been powerful incentives for committee members to keep a lid on the Climategate affair. Environmentalism has been a key part of David Cameron's rebranding of the Conservative Party, and it has undoubtedly been a factor in his ability to win over swing voters. But as well as this, green issues are a key part of the Liberal Democrats' policy platform, and an area in which the party is very mindful of losing votes to the Greens. In these circumstances the Conservative whips would almost certainly have been unhappy with any committee member who voted in favour of an amendment that damaged the Prime Minister's reputation *and* relations inside the coalition. This is, as I have said, pure speculation, but it would explain the decision to strike out the Stringer amendment and issue such a mealy mouthed report in the face of overwhelming evidence that the inquiries were whitewashes.

If this speculation is correct, then it is presumably superfluous to look to government for help. Ministers appear to have restricted themselves to expressions of belief in the integrity of British science and relief that the exonerations were issued, although it is possible that they were actually more closely involved. One possibility is that Beddington's involvement in the inquiries may have come at the behest of ministers, although we have no evidence that this was the case. In some ways, however, this is not a particularly relevant consideration. If ministers have not involved themselves in an issue considered important enough to need a parliamentary inquiry then they are clearly giving miscreants a free ride. If they have actively involved themselves in the coverups then the situation is certainly no better.

THE UNACCOUNTABILITY OF THE INSTITUTIONS

The tale of Climategate and its aftermath is not an edifying one. As we look back over the ten years of this story, the impression we get is of a wave of dishonesty, a public sector that will spin and lie, and mislead and lie, and distort and lie, and lie again. If one lie fails then another lie is issued and if that fails then they simply lie again. And all this happens without fear of the consequences: everyone involved appears quite certain that their mendacity will go unpunished no matter what.

What is worse, it appears that public servants can also *break the law* with impunity. We have already seen that the public servants responsible for FOI legislation in the UK managed to frame the law in a way that made it virtually impossible for any breach ever to be prosecuted. It appears unlikely that this situation will be changed in the near future. Meanwhile, the failure of two inquiries to investigate breaches of the Act – Russell because he thought the allegations too serious and the Information Commissioner because of the six-month statute of limitations – appears to have represented carte blanche for UEA to continue blithely as if nothing had happened. Despite the emails showing, apparently incontrovertibly, that FOI laws were flouted with the full knowledge of senior figures in university, there have been almost no discernible repercussions for anyone involved. It seems that criminal behaviour is no bar to employment at the highest levels of a university. And this appears to be accepted by everyone involved or who might be in a position to do something about it, from the governing body of the university to parliamentarians, ministers and the media. Climategate was, as Fred Pearce put it, a tragedy born of misunderstood motives. The response to it was an extraordinary failure of the institutions and of the people who are paid to protect the public interest – a failure of honesty, a failure of diligence, a failure of integrity. Their failure to seek the truth and to speak the truth condemns them utterly.

TIMELINE

Date	Russell	Parliament	Oxburgh	Penn State
Nov '09		Climategate		Launch
Dec '09	Announced			
Jan '10				
Feb '10		Announced		Interview
				Inq. report
Mar '10	Launch	Hearings	Announced	
Apr '10		Report	Launch	
May '10			Report	
Jun '10				
Jul '10	Report			Inv. report
Aug '10				
Sep '10				
Oct '10		Oxburgh hearing		
⋮		Russell hearing		
Jan '11				
⋮		Second report		
Nov '11		Climategate 2		

BIBLIOGRAPHY

PRINCIPAL SOURCES

BHE — Sir Brian Hoskins' emails. The emails are available at: http://bit.ly/wlZmSm, archived at: http://www.webcitation.org/5zXJ61DrJ. Some names have been redacted, but have since been disclosed. See: http://www.webcitation.org/5zXJIj3tl.

CG1, CG2 — The Climategate emails are listed by email number, with a prefix of CG1 or CG2 to indicate whether the reference is to the 2009 or 2011 disclosures. Since many of the emails are threads, rather than single messages, the dates given may refer to the date of the thread or the date of the particular message.

DHE, DHA — David Hand's emails and attachments. The emails are available at: http://bit.ly/zuLzQe, archived at: http://www.webcitation.org/65S1Qc0Qs. The attachments are available at: http://bit.ly/wEfI6M, archived at: http://www.webcitation.org/65S1Kazih.

HSI — Montford, A. *The Hockey Stick Illusion*. Stacey International, 2010.

RR — Report of the Independent Climate E-mails Review (The Russell report). The report is available at http://www.cce-review.org/pdf/FINALREPORT.pdf and has been archived at http://www.webcitation.org/66BssaW0c.

STCE I — Evidence submitted to the first House of Commons Science and Technology Committee hearings. Pages numbers are given as per the published evidence, for example Ev 10. The evidence is at: http://www.publications.parliament.uk/pa/cm200910/cmselect/cmsctech/387/387ii.pdf (archived at http://www.webcitation.org/66EzJwlO1)

STCR I — Report of the first Science and Technology Committee. The report is available at : http://www.publications.parliament.uk/pa/cm200910/cmselect/cmsctech/387/387i.pdf (archived at: http://www.webcitation.org/66EzFvrY2)

STCE II   References to evidence submitted to the second House of Commons Science and Technology Committee hearings are given by STCE II. The evidence and report were published together in a document available at: http://www.publications.parliament.uk/pa/cm201011/cmselect/cmsctech/444/444.pdf (archived at: http://www.webcitation.org/66Ey92FV7), with additional written evidence at: http://www.publications.parliament.uk/pa/cm201011/cmselect/cmsctech/444/444vw.pdf (archived at: http://www.webcitation.org/66EyN9j0B).

CITATIONS

1. Soon W, Baliunas S. Proxy climatic and environmental changes of the past 1000 years. Climate Research. 2003;23:89–110.

2. CG1: 1047388489; 11 March 2003.

3. CG2: 2272; 16 April 2003.

4. CG2: 0332; 22 April 2003.

5. CG2: 3052; 24 April 2003.

6. CG2: 1185; 24 April 2003.

7. CG1: 1051190249; 24 April 2003.

8. CG1: 1051202354; 24 April 2003.

9. CG2: 1430; 28 April 2003.

10. CG2: 4132; 28 April 2003.

11. CG2: 4808; 16 May 2003.

12. Goodess C. Stormy times for climate research. SGR Newsletter. 2003;28:(online). Available at: http://www.sgr.org.uk/resources/stormy-times-climate-research. Archived at: http://www.webcitation.org/63gPXI6BP.

13. CG1: 1057944829; 11 July 2003.

14. de Freitas C. Email to the author; 2011.

15. CG1: 1057941657; 11 July 2003.

16. CG2: 2280; 24 July 2003.

17. CG2: 5025; 25 July 2003.

18. CG2: 4321; 22 September 2003.

19. CG1: 1067194064; 26 October 2003.

20. Osborn T, Briffa K, Jones P. Note on paper by McIntyre and McKitrick in *Energy and Environment*. Climatic Research Unit website; November 2003. Archived at: http://replay.waybackmachine.org/20040410001632/http://www.cru.uea.ac.uk/~timo/paleo/.

21. CG1: 1067596623; 31 October 2003.

22. CG1: 1098472400; 22 October 2004.

23. CG2: 3259; 22 October 2004.

24. CG1: 1106322460; 20 January 2005.

25. CG1: 1107899057; 4 February 2005.

26. CG1: 1104855751; 30 December 2004.

27. CG1: 1121392136; 14 July 2005.

28. CG1: 1116902771; 23 May 2005.

29. CG1: 1104893567; 4 January 2005.

30. CG1: 1120528403; 4 July 2005.

31. CG1: 1132094873; 15 November 2005.

32. CG1: 1133360497; 30 November 2005.

33. Holland D. Submission to the Independent Climate Change Email Review; 25 February 2010. Available at: http://homepages.tesco.net/~kate-and-david/temp/Holland%20ICCER%20Submission%20Rev1.pdf. Archived at: http://www.webcitation.org/5xjthEuXZ. The version at these sites is actually a later revision of the document, but according to the revision record at the end of the document, the differences are minor.

34. Manning M. Deadlines for literature cited in the Working Group I Fourth Assessment Report; 1 June 2005. Archived at: http://classic-web.archive.org/web/20060207153611/ipcc-wg1.ucar.edu/wg1/PublicationDeadlines.pdf.

35. CG2: 4201; 13 December 2005.

36. CG1: 1138042050; 23 January 2006.

37. CG1: 1139591144; 10 February 2006.

38. CG1: 1141180962; 28 February 2006. Holland quoted this at paragraph 44 of his submission to Russell.

39. CG1: 1139845689; 13 February 2006.

40. CG2: 0790; 28 February 2006.

41. CG1: 1138995069; 3 February 2006.

42. HSI; Chapter 9.

43. Jansen, E, Overpeck J, Briffa KR, Duplessy JC, Joos F, et al. Paleoclimate. In: Climate Change 2007: The Physical Science Basis. Contribution of Working Group I to the Fourth Assessment Report of the Intergovernmental Panel on Climate Change (Second Order Draft); 2006. Available at: http://pds.lib.harvard.edu/pds/view/7768990.

44. HSI; Chapter 8.

45. CG1: 1151094928; 23 June 2006.

46. CG1: 1148339153; 28 May 2006.

47. Briffa KR, Osborn TJ. Response to specific questions raised by Professor Geoffrey Boulton in his letter of 6 May 2010, in his role as a member of the Muir-Russell review team; 19 May 2010. Available at: http://www.cce-review.org/evidence/6%20May%20Briffa%20Osborn%20response.pdf. Archived at: http://www.webcitation.org/62nC3Jzhx.

48. Manning M. Guidelines for inclusion of recent scientific literature in the Working Group I Fourth Assessment Report; 1 July 2006. Available at: http://web.archive.org/web/20070206012931/ipcc-wg1.ucar.edu/wg1/docs/PublicationDeadlines_2006-07-01.pdf.

49. IPCC. Fourth Assessment Report, Second Order Draft review comments; 2006. The comment discussed is number 11-7.

50. CG1: 1150923423; 21 June 2006.

51. CG1: 1147982305; 18 May 2006.

52. CG1: 1153470204; 21 July 2006.

53. HSI; Chapter 12.

54. CG1: 1155402164; 12 August 2006.

55. Jansen, E, Overpeck J, Briffa KR, Duplessy JC, Joos F, et al. Paleoclimate. In: Solomon, S, Qin D, Manning M, Chen Z, Marquis M, et al., editors. Climate Change 2007: The Physical Basis. Cambridge University Press; 2007. .

56. Mitchell J. Review of IPCC Fourth Assessment Report Chapter 6; 8 December 2006. As noted in the text, the original date of the report is 8 December 2007, but at some point afterwards it has been amended in another hand to 2006. A facsimile is available at: http://www.bishop-hill.net/inquiries-files/20061208Mitchell%27s%20Review%20Editor%20Signoff.pdf. Archived at: http://www.webcitation.org/62nIwSIBg.

57. Solomon S, Mitchell J. Email correspondence related to David Holland's FOI request; 14 March 2008. Available at: http://www.ventalize.org.uk/Climate%20Change/Climategate/Solomon_20080314.pdf. Archived at http://www.webcitation.org/67hMb54FO.

58. Mitchell J. Letter to David Holland; 27 March 2008. Quoted at: http://climateaudit.org/2008/05/02/no-working-papers-no-correspondence/. Archived at: http://www.webcitation.org/5zX47HPqx.

59. CG1: 1210341221; 9 May 2008.

60. McKitrick R. Comment on IPCC review process; *Climate Audit* blog; 11 March 2011. Available at: http://climateaudit.org/2011/03/10/what-did-penn-state-know/. Archived at: http://www.webcitation.org/5xAuNmNlJ.

61. McIntyre S. Wahl and Ammann 2007 and IPCC deadlines. *Climate Audit* blog; 25 May 2008. Available at: http://climateaudit.org/2008/05/25/wahl-and-ammann-2007-and-ipcc-deadlines/. Archived at: http://www.webcitation.org/5wuNU63an.

62. CG1: 1212156886; 30 May 2008.

63. CG1: 1212009215; 28 May 2008.

64. CG1: 1212063122; 29 May 2008.

65. McIntyre S. Wahl transcript excerpt. *Climate Audit* blog; 8 March 2011. Available at: http://climateaudit.org/2011/03/08/wahl-transcript-excerpt/. Archived at: http://www.webcitation.org/62nL6cSIf.

66. CG2: 2526; 4 June 2008.

67. CG2: 1299; 21 February 2005.

68. CG1: 1106338806; 21 January 2005.

69. CG1: 1107454306; 2 February 2005.

70. Jones P, Groisman PY, Coughlan M, Plummer N, Wang W, Karl T. Assessment of urbanization effects in time series of surface air temperature over land. *Nature*. 1990;347:169–172.

71. McIntyre S. Phil Jones and the great leap forward; *Climate Audit* blog; 26 February 2007. Available at: http://climateaudit.org/2007/02/26/phil-jones-and-the-great-leap-forward/. Archived at :http://www.webcitation.org/5z6tstcrB.

72. CG1: 1177158252; 20 April 2007.

73. CG1: 177534709; 25 April 2007.

74. Zhongwei Y, Chi Y, Jones P. Influence of inhomogeneity on the estimation of mean and extreme temperature trends in Beijing and Shanghai. *Advances in Atmospheric Sciences*. 2001;18(3):309–322.

75. CG1: 118234247; 20 June 2007.

76. Keenan D. Remarks related to my exposé, 'The fraud allegation against some climatic research of Wei-Chyung Wang'; 2007–2010. Available at: http://informath. org/apprise/a5620.htm. Archived at: http://www.webcitation.org/67g2fMEAB.

77. CG1: 1219239172; 20 August 2008.

78. CG1: 1228330629; 3 December 2008.

79. Climatic Research Unit. CRU data availability; 2009. Available at: http://www. cru.uea.ac.uk/cru/data/availability/ Archived at: http://www.webcitation.org/ 5w31clOcP.

80. HSI; Chapter 10.

81. Briffa KR, Shishov VV, Melvin TM, Vaganov EA, Grudd H, Hantemirov RM, et al. Trends in recent temperature and radial tree growth spanning 2000 years across northwest Eurasia. Phil Trans R Soc B. 2008;363:2269–82.

82. McIntyre S. YAD06 – the most influential tree in the world; *Climate Audit* blog; 30 September 2009. Available at: http://climateaudit.org/2009/09/30/yamal-the-forest-and-the-trees/. Archived at: http://www.webcitation.org/67g7Vva9I.

83. Anonymous. Hey Ya! (mal); *RealClimate* blog; 30 September 2009. Available at: http://www.realclimate.org/index.php/archives/2009/09/hey-ya-mal/. Archived at: http://www.webcitation.org/67X21BmvP.

84. Briffa K, Melvin TM. Examining the validity of the published RCS Yamal tree-ring chronology; 28 October 2009. Available at: http://www.cru.uea. ac.uk/cru/people/briffa/yamal2009/. Archived at: http://www.webcitation.org/ 67X2C9QQX.

85. McIntyre S. Miracles and strip bark standardization; *Climate Audit* blog; 16 November 2009. Available at http://climateaudit.org/2009/11/16/luckman-at-the-canadian-society-for-petroleum-geologists.

86. RC. Comment left at *Climate Audit*; 17 November 2009. The comment can still be seen at http://climateaudit.org/2009/11/16/luckman-at-the-canadian-society-for-petroleum-geologists/.

87. Schmidt G. One year later; *RealClimate* blog; 22 November 2010. Available at: http://www.realclimate.org/index.php/archives/2010/11/one-year-later/.

88. 'RC'. Comments left at *Watts Up With That?*, *Climate Skeptic* and *The Airvent*; 17 November 2009. The comment at *Watts Up With That?* was noticed by moderators and was removed. The other comments are still in place and can be seen at http:// www.climate-skeptic.com/2009/11/bummer-i-didnt-make-the-list.html and http: //noconsensus.wordpress.com/2009/11/13/open-letter respectively.

89. Watts A. Email to the author; 1 May 2012.

90. Dennis P. Comment at *Bishop Hill* blog; 5 February 2010, 08:33. Available at http://bishophill.squarespace.com/blog/2010/2/4/a-mention-in-the-guardian.html. Archived at: http://www.webcitation.org/65RmOVr1K.

91. Mosher S. Comment at *The Blackboard* blog; 19 November 2009. Available at: http://rankexploits.com/musings/2009/giss-october-anomaly-same-as-september/#comment-23722. Archived at: http://www.webcitation.org/67MAPoQXM.

92. Schmidt G. Email to Lucia Liljegren; 19 Nov 2009 15:48:21 -0500. Reproduced at http://climateaudit.org/2010/01/12/the-mosher-timeline/. Archived at: http://www.webcitation.org/65RmaVbVa.

93. Thomson D. Comment at *Climate Audit* blog; 19 November 2009, 19:41. Available at http://climateaudit.org/2009/11/19/cru-correspondence/. Archived at: http://www.webcitation.org/65RmoU8Bo.

94. Pearce F. *The Climate Files*. Guardian Books; 2010.

95. Wishart I. Climate centre hacked. TGIF edition; 20 November 2009. Available at: http://issuu.com/iwishart/docs/tgif20nov09?viewMode=magazine. A PDF copy is archived at: http://www.webcitation.org/5xt8Iczlz.

96. CG2: 3451; 16 November 1999.

97. Anonymous. The CRU hack. *RealClimate* blog; 20 November 2009. Available at: http://www.realclimate.org/index.php/archives/2009/11/the-cru-hack/. Archived at: http://www.webcitation.org/5xt9avGo7.

98. Hickman L, Randerson J. Climate sceptics claim leaked emails are evidence of collusion among scientists. *Guardian* website; 20 November 2009. Available at: http://www.guardian.co.uk/environment/2009/nov/20/climate-sceptics-hackers-leaked-emails. Archived at: http://www.webcitation.org/5wW59AErd.

99. Bolt A. Climategate: warmist conspiracy exposed? *Herald Sun* website; 20 November 2009. Available at: http://www.webcitation.org/5wW7Ytt1L.

100. Delingpole J. Climategate: the final nail in the coffin of 'anthropogenic global warming'? James Delingpole's blog; 20 November 2009. Available at: http://blogs.telegraph.co.uk/news/jamesdelingpole/100017393/climategate-the-final-nail-in-the-coffin-of-anthropogenic-global-warming/. Archived at:http://www.webcitation.org/5wW7waoWG.

101. Revkin A. Hacked e-mail is new fodder for climate dispute. *New York Times DotEarth*; 20 November 2009. Available at: http://www.nytimes.com/2009/11/21/science/earth/21climate.html?_r=1%. Archived at: http://www.webcitation.org/5wW94RV0I.

102. Eilperin J. In the trenches on climate change, hostility among foes. *Washington Post* website; 22 November 2009. Available at: http://www.washingtonpost.com/wp-dyn/content/article/2009/11/21/AR2009112102186.html. Archived at: http://www.webcitation.org/5wW8CtvJo.

103. Anonymous. Climate skeptics see 'smoking gun' in researchers' leaked e-mails; *Fox News*; 21 November 2009. Available at: http://www.foxnews.com/scitech/2009/11/21/climate-skeptics-smoking-gun-researchers-leaked-e-mails/. Archived at: http://www.webcitation.org/65RnDwZXu.

104. Revkin A. Private climate conversations on display. *New York Times DotEarth* blog; 21 November 2009. Available at: http://dotearth.blogs.nytimes.com/2009/11/20/private-climate-conversations-on-display/. Archived at: http://www.webcitation.org/5wW8auCux.

105. Pierrehumbert R. Email to Andrew Revkin. Reproduced at *DotEarth* blog; 22 November 2009. Available at: http://dotearth.blogs.nytimes.com/2009/11/22/your-dot-on-science-and-cyber-terrorism/. Archived at: http://www.webcitation.org/5wWA2c4tS.

106. Weinberg K. Nigel Lawson: Thatcher's chancellor takes on the planet alone; *Daily Telegraph*; 21 November 2009. Available at: http://www.telegraph.co.uk/earth/copenhagen-climate-change-confe/6618848/Nigel-Lawson-Thatchers-Chancellor-takes-on-the-planet-alone.html. Archived at: http://www.webcitation.org/5xtAPaXbn.

107. Moore M. Lord Lawson calls for public inquiry into UEA global warming data 'manipulation'; *Daily Telegraph*; 23 November 2009. Available at: http://www.telegraph.co.uk/earth/environment/globalwarming/6634282/Lord-Lawson-calls-for-public-inquiry-into-UEA-global-warming-data-manipulation.html. Archived at: http://www.webcitation.org/5xtACoZlB.

108. University of East Anglia Press Office. CRU update 1; 23 November 2009. Available at: http://www.uea.ac.uk/mac/comm/media/press/2009/nov/CRU-update. Archived at: http://www.webcitation.org/5xsrkNTwH.

109. University of East Anglia press office. CRU update 2; 24 November 2009. Available at: http://www.uea.ac.uk/mac/comm/media/press/2009/nov/CRUupdate. Archived at: http://www.webcitation.org/5w30Hx9B4.

110. Channel Four News. News segment on Climategate; 24 November 2009. Available at: http://link.brightcove.com/services/player/bcpid1184614595?bctid=52457182001.

111. Editorial. Global warming with the lid off; *Wall Street Journal*; 24 November 2009. Available at: http://online.wsj.com/article/SB10001424052748704888404574547730924988354.html. Archived at: http://www.webcitation.org/62U9TxM2d.

112. Crook C. Obama offers cuts at Copenhagen; *Financial Times* blog; 25 November 2009. Available at: http://blogs.ft.com/crookblog/2009/11/obama-offers-cuts-at-copenhagen/#axzz1ayiyCzNv. Archived at: http://www.webcitation.org/62U9jdAjT.

113. Spencer R. Global warming's blue dress moment? the CRU email hack scandal. Dr Roy Spencer's blog; 20 November 2009. Available at:http://www.drroyspencer.com/2009/11/global-warmings-blue-dress-moment-the-cru-email-hack-scandal/. Archived at: http://www.webcitation.org/5wWHnXgH9.

114. Zorita E. Why I think that Michael Mann, Phil Jones and Stefan Rahmstorf should be barred from the IPCC process. Personal website; 28 November 2009. Zorita's article is no longer available on his website, but can still be seen on the website of Roger Pielke Jnr at: http://rogerpielkejr.blogspot.com/2009/11/eduardo-zorita-on-climategate.html. Archived at: http://www.webcitation.org/5wWI8ewzG.

115. Koutsoyiannis D. Beware saviors! *Roger Pielke Snr* blog; 24 November 2010. Available at: http://pielkeclimatesci.wordpress.com/2009/11/24/beware-saviors-by-demetris-koutsoyiannis/. Archived at: http://www.webcitation.org/65RnXrTCe.

116. Munro N. Climate scientist: time for more transparency. Interview with Judith Curry. *National Review* website; 2 December 2009. Available at: http://insiderinterviews.nationaljournal.com/2009/12/email-controversy-divides.php. Archived at: http://www.webcitation.org/5wXU01vFy.

117. Hayward S. Climate scientist to Revkin: 'we can no longer trust you' to carry water for us. Ashbrook Center website; 6 December 2009. Available at: http://nlt.ashbrook.org/2009/12/climate-scientist-to-revkin-we-can-lo-longer-trust-you-to-carry-water-for-us.php. Archived at: http://www.webcitation.org/5wWVrEdGr.

118. Steig E. Who you gonna call? *RealClimate* blog; 5 December 2009. Available at: http://www.realclimate.org/index.php/archives/2009/12/who-you-gonna-call/.

119. Halpin T. Is Russia behind the Climategate hackers? *Times* website; 7 December 2009. Available at: http://www.timesonline.co.uk/tol/news/environment/article6946385.ece. Archived at: http://www.webcitation.org/5wWXFgflG.

120. McKie R, Vidal J. Break-in targets climate scientist. *Guardian* website; 6 December 2009. Available at: http://www.guardian.co.uk/science/2009/dec/06/break-in-targets-climate-scientist. Archived at: http://www.webcitation.org/5wWX2BBl9.

121. CG1: 0938018124; 22 September 1999.

122. CG2: 0697; 23 September 1999.

123. McIntyre S. IPCC and the 'Trick'; *Climate Audit* blog; 10 December 2009. Available at: http://climateaudit.org/2009/12/10/ipcc-and-the-trick/. Archived at: http://www.webcitation.org/5xtB7WDGl.

124. Rose D. Special Investigation: Climate change emails row deepens as Russians admit they *did* come from their Siberian server. *Daily Mail*; 13 December 2009. Available at: http://www.dailymail.co.uk/news/article-1235395/SPECIAL-INVESTIGATION-Climate-change-emails-row-deepens--Russians-admit-DID-send-them.html. Archived at: http://www.webcitation.org/681AOHkbb.

125. Leake J. Climate data dumped. *Sunday Times* website; 29 November 2009. Available at: http://www.timesonline.co.uk/tol/news/environment/article6936328.ece. Archived at: http://www.webcitation.org/5wXdGItW1.

126. University of East Anglia press release. CRU update 3; 1 December 2009. Available at: http://www.uea.ac.uk/mac/comm/media/press/2009/dec/CRUphiljones. Archived at: http://www.webcitation.org/5wXNEFciV.

127. University of East Anglia press release. Sir Muir Russell to head the independent review into the allegations against the Climatic Research Unit (CRU); 3 December 2009. Available at: http://www.uea.ac.uk/mac/comm/media/press/2009/dec/CRUreview. Archived at: http://www.webcitation.org/5wXN4NrbL.

128. Acton E. Letter to Phil Willis; 10 December 2009. Available at: http://www.parliament.uk/documents/upload/091210-uea-vice-chancellor-letter.pdf. Archived at: http://www.webcitation.org/5xtC8Iodd.

129. Delingpole J. Wow! UK parliamentary investigation into Climategate may not be a whitewash; James Delingpole's blog; 22 January 2009. The original article was not published online, but was quoted that day at James Delingpole's blog: http://blogs.telegraph.co.uk/news/jamesdelingpole/100023449/wow-uk-parliamentary-investigation-into-climategate-may-not-be-a-whitewash/. Archived at: http://www.webcitation.org/5wXek5pOW.

130. Russell M. Minutes of meeting with UEA officials; 18 December 2009. Available at: http://www.cce-review.org/pdf/MR%2018%20Dec%20final%20Registrar%20etc.pdf. Archived at: http://www.webcitation.org/5x57Lk7mh.

131. McIntyre S. Comment on CRUTEM. *Climate Audit* blog; 28 July 2009. Available at: http://climateaudit.org/2009/07/25/a-mole/. Archived at: http://www.webcitation.org/62qAJCd0U.

132. Editorial. Climatologists under pressure. *Nature*. 2009;462:545.

133. CRI English website. Interview with Philip Campbell; 3 December 2009. Audio at: http://english.cri.cn/7146/2009/12/03/1901s533264.htm. Partial transcript available at: http://www.bishop-hill.net/blog/2010/2/11/russell-review-under-way.html. Archived at: http://www.webcitation.org/5wXk88GqA.

134. Clarke T. 'Climate-gate' review member resigns; *Channel Four News* website; 11 February 2010. Available at: http://www.channel4.com/news/articles/science_technology/aposclimategateapos+review+member+resigns/3536642.html. Archived at: http://www.webcitation.org/5wXkJmXfy.

135. Boulton G. Remarks made to meeting of the Foundation for Science and Technology. FST website; 29 October 2009. Available at: http://www.foundation.org.uk/events/pdf/20091029_Summary.pdf. Archived at: http://www.webcitation.org/5wXgXTXQN.

136. Boulton G. Lecture to the Glasgow Centre for Population Health; 29 January 2008. Transcript available at: http://www.gcph.co.uk/assets/0000/0421/Geoffrey_Boulton_Transcript.pdf. Archived at: http://www.webcitation.org/5xsyxF3tV.

137. Met Office. Statement from the UK science community; 10 December 2009. Archived at: http://tna.europarchive.org/20091209145453/http:/www.metoffice.gov.uk/climatechange/news/latest/uk-science-statement.html.

138. Royal Society of Edinburgh. Briefing paper 09-05; December 2009. Available at: http://www.rse.org.uk/govt_responses/2009/b09_05.pdf. Archived at: http://www.webcitation.org/5x6QFTTTG.

139. Campbell R. Submission of evidence to the Russell review; 16 February 2010. Available at: http://www.cce-review.org/evidence/Campbell.pdf. Archived at: http://www.webcitation.org/5wXiHGuPH.

140. McIntyre S. Partial transcript of inquiry press conference. *Climate Audit* blog; 16 February 2010. Available at: http://climateaudit.org/2010/02/16/partial-transcript-of-inquiry-press-conference/. Archived at: http://www.webcitation.org/5zCJDVgrG.

141. Webster B. UN must investigate warming 'bias', says former climate chief. *Times* website; 15 December 2010. Available at: http://www.timesonline.co.uk/tol/news/environment/article7026932.ece. Archived at: http://www.webcitation.org/5x6OzLF4T.

142. Climate Change Emails Review press release. Allegations of bias against review member rejected; 15 February 2010. Available at: http://www.cce-review.org/News.php. Archived at: http://www.webcitation.org/5wXnWE3bX.

143. McIntyre S. Who wrote the 'issues paper'? *Climate Audit* blog; 15 February 2010. Available at: http://climateaudit.org/2010/02/15/who-wrote-the-issues-paper/. Archived at: http://www.webcitation.org/5x6WHW4cw.

144. Climate Change Emails Review. Workplan; 11 February 2010. Available at: http://www.cce-review.org/Workplan.php. Archived at: http://www.webcitation.org/5we8kD6Bs.

145. Climate Change Emails Review. Issues for examination; 11 February 2010. Available at: http://www.cce-review.org/pdf/CCER%20ISSUES%20FOR%20EXAMINATION%20FINAL.pdf. Archived at: http://www.webcitation.org/5we8xnI9s.

146. University of East Anglia Press Office. New scientific assessment of climatic research publications announced; 11 February 2010. Available at: http://www.uea.ac.uk/mac/comm/media/press/CRUstatements/

Newscientificassessmentofclimaticresearchpublicationsannounced. Archived at: http://www.webcitation.org/5wzTLgYhL.

147. McIntyre S. Muir Russell and the team. *Climate Audit* blog; 11 February 2010. Available at: http://climateaudit.org/2010/02/11/a-muir-russell-avatar/. Archived at:http://www.webcitation.org/5we9RL8ZT.

148. Climate Change Emails Team. Minutes of meeting with UEA IT staff; 18 December 2009. Available at: http://www.cce-review.org/pdf/MR%2018%20Dec%20final%20IT%20Personnel.pdf. Archived at: http://www.webcitation.org/5xcP9KYBr.

149. Hudson P. 'Climategate' – CRU hacked into and its implications; BBC *Paul Hudson's blog*; 23 November 2009. Available at: http://www.bbc.co.uk/blogs/paulhudson/2009/11/climategate-cru-hacked-into-an.shtml. Archived at: http://www.webcitation.org/5xAyt1JUV.

150. Hudson P. 'Climategate' - What next?; BBC *Paul Hudson's blog*; 24 November 2009. Available at: http://www.bbc.co.uk/blogs/paulhudson/2009/11/climategate-what-next.shtml. Archived at: http://www.webcitation.org/5xAyy8bCK.

151. Russell M. Minutes of meeting with Phil Jones; 18 December 2009. Available at: http://www.cce-review.org/pdf/MR%2018%20Dec%20final%20P%20Jones.pdf. Archived at: http://www.webcitation.org/5x586yVPG.

152. CCE emails team. Minutes of meeting with Keith Briffa; 18 December 2009. Available at: http://www.cce-review.org/pdf/MR%2018%20Dec%20final%20K%20Briffa.pdf. Archived at: http://www.webcitation.org/60FMrq4TH.

153. CG1: 1228922050; 10 December 2008.

154. Rosenbaum M. Hacked climate e-mails and FOI. BBC *Open Secrets* blog; 23 November 2009. Available at: http://www.bbc.co.uk/blogs/opensecrets/2009/11/hacked_climate_emails_and_foi.html. Archived at: http://www.webcitation.org/5zX7DCkWU.

155. Campaign for Freedom of Information. Time limit for prosecution of offences under section 77 of the FOI Act; 29 January 2010. Available at: http://www.cfoi.org.uk/fois77offence290110.html. Archived at: http://www.webcitation.org/5xsvNDPpb.

156. Russell inquiry team. Minutes of meeting; 12 January 2010. Available at: http://www.cce-review.org/pdf/CCER%20initial%20meeting%2012%20Jan.pdf. Archived at: http://www.webcitation.org/5x5BrP46G.

157. Climate Change Emails Team. Notes of meeting with Mick Gorrill and David Clancy; 27 January 2010. Available at: http://www.cce-review.org/evidence/UEA-CRU_IV2_ICO_270110FIN2.pdf. Archived at: http://www.webcitation.org/5xBA2TIs5.

158. Climate Change Emails Team. Notes of an Interview with Prof. Phil Jones and Prof. Keith Briffa; 27 January 2010. Available at: http://www.cce-review.org/evidence/UEA-CRU_IV4_Briffa%20and%20Jones_v6_1.pdf. Archived at: http://www.webcitation.org/5xBC6CG07.

159. Information Commissioner's Office. Correspondence related to statute of limitations; January–February 2010. Available at: http://www. whatdotheyknow.com/request/28506/response/74321/attach/3/Copy%20of% 20information%20in%20response%20to%20your%20request.PDF.pdf. Archived at: http://www.webcitation.org/5xswp39v8.

160. University of East Anglia. Memorandum issued to Science and Technology Committee; 25 February 2010. Available at: http://www.publications.parliament.uk/pa/ cm200910/cmselect/cmsctech/memo/climatedata/uc0002.pdf. Archived at: http://www.webcitation.org/5xstN68HJ.

161. University of East Anglia. Correspondence between University of East Anglia and the Information Commissioner's Office; 26 February 2010. Available at: http://www.uea.ac.uk/mac/comm/media/press/CRUstatements/ ICOcorrespondence. Archived at: http://www.webcitation.org/5xcXtOE7B.

162. Webster B. University 'tried to mislead MPS on climate change e-mails'. *Times* website; 27 February 2010. Available at: http://www.timesonline.co.uk/tol/ news/environment/article7043566.ece. Archived at: http://www.webcitation. org/5xsviCao6.

163. McIntyre S. Hard to imagine more cogent prima facie evidence; *Climate Audit* blog; 26 February 2010. Available at: http://climateaudit.org/2010/02/26/ hard-to-imagine-more-cogent-prima-facie-evidence/. Archived at: http://www. webcitation.org/5zXC8DQ6d.

164. Penn State Press Office. University reviewing recent reports on climate information; 25 November 2009. Available at: http://www.ems.psu.edu/sites/default/ files/u5/Mann_Public_Statement.pdf. Archived at: http://www.webcitation.org/ 66NjBzfhe.

165. Nichols L. PSU investigates 'Climategate'; 30 November 2009. Available at: http://www.collegian.psu.edu/archive/2009/11/30/psu_investigates_ climategate.aspx. Achived at: http://www.webcitation.org/63xOQHVKH.

166. Nichols L. PSU panel to review all 'Climategate' e-mails; 9 December 2009. Available at: http://www.collegian.psu.edu/archive/2009/12/09/psu_panel_to_ review_all_climat.aspx. Archived at: .

167. Foley HC, Scaroni AW, Yekel CA. RA-10 Inquiry Report: Concerning the Allegations of Research Misconduct; 3 February 2010. Available at: http://www.research.psu. edu/orp/Findings_Mann_Inquiry.pdf. Archived at: http://www.webcitation.org/ 68wCbQOFn.

168. Freedman A. Scientist: Consensus withstands climate e-mail flap; 1 December 2009. Available at: http://voices.washingtonpost.com/capitalweathergang/ 2009/12/gerald_north_interview.html. Archived at: http://www.webcitation.org/ 64Ppktjod.

169. Kennedy D. Editorial: silly season on the Hill. *Science*. 2005;309:1301. Available at: http://www.sciencemag.org/content/309/5739/1301.full.pdf. Archived at: http://www.webcitation.org/63ySFyDsi.

170. Lindzen R, North G. The Great Climate Change Debate: Where is the Physical Science? Rice University; 27 January 2010. Audio available at: http://wmdp.rice.edu/Centers/CSES/ClimateChg-27Jan10/ClimateChg-27Jan10.mp3.

171. Penn State University. Policy RA10 Handling inquiries/investigations into questions of ethics in research and in other scholarly activities; 18 May 2007. Available at: http://guru.psu.edu/policies/RA10.html. Archived at: http://www.webcitation.org/63yW9fzOk.

172. Boyle C. Mann inquiry concludes, board to release findings; 1 February 2010. Available at: http://www.collegian.psu.edu/archive/2010/02/01/mann_inquiry_concludes_board_t.aspx. Archived at: http://www.webcitation.org/64RURLiAA.

173. Martin C. Climategate meets the law: Senator Inhofe to ask for DOJ investigation; *PJ Media*; 23 February 2010. Available at: http://pjmedia.com/blog/climategate-and-the-law-senator-inhofe-to-ask-for-congressional-criminal-investigation-pajamas-mediapjtv-exclusive/. Archived at: http://www.webcitation.org/63yX19GVK.

174. STCE I; Ev 169.

175. STCE; Ev 144.

176. STCE I; Ev 19.

177. STCE I; Ev 5.

178. STCR; p. 54.

179. STCE I; Ev 149.

180. Jones P. Review of Mann *et al.* manuscript; 1 November 2008. Available at: http://www.climate-gate.org/cru/documents/review_mannetal.doc. Archived at: http://www.webcitation.org/5zDRNvj5q.

181. Jones P. Review of Schmidt *et al.* 2009; 22 June 2008. Available at: http://www.climate-gate.org/cru/documents/review_schmidt.doc. Archived at: http://www.webcitation.org/5zDQzAcO0.

182. Jones P. Review of Wahl and Ammann 2007; 18 May 2005. Available at: http://www.climate-gate.org/cru/documents/Review%20of%20Wahl&Amman.doc. Archived at: http://www.webcitation.org/5zDRqIU09.

183. STCE I; Ev 21.

184. STCE I; Ev 143.

185. STCE I; Ev 34.

186. STCE I; Ev 181.

187. STCR I; p. 3.

188. HSI; Chapter 3.

189. STCE I; Ev 20.

190. STCE I; Ev 29.

191. STCE I; Ev 2.

192. STCE I; Ev 41.

193. STCE I; Ev 42.

194. Furedi F. 'Climategate': what a pointless investigation; 31 March 2010. Available at: http://www.frankfuredi.com/index.php/site/article/382/. Archived at: http://www.webcitation.org/6089pWyU9.

195. House of Commons Science and Technology Committee. Press release: Climate science must become more transparent say MPs; 31 March 2010. Available at: http://www.parliament.uk/business/committees/committees-archive/science-technology/s-t-pn32-100331/. Archived at: http://www.webcitation.org/6082ZPUnf.

196. Willis P. Interview with BBC Radio 4 *Today* programme; 31 March 2010. Available at: http://news.bbc.co.uk/1/hi/8595483.stm.

197. Jones P. Letter to Swedish Meteorological and Hydrological Insitute; 30 November2009. Available at: http://wattsupwiththat.files.wordpress.com/2010/03/doc111209.pdf. Archived at: http://www.webcitation.org/67gK9GCQn.

198. Flarup M. Letter to Phil Jones; 21 December 2009. Available at: http://wattsupwiththat.files.wordpress.com/2010/03/request_from_professor_phil_jones_regarding_the_release_of_data_from_the_hadcrut_dataset__dnr_smhi_.pdf. Archived at: http://www.webcitation.org/632gi32op.

199. STCE I; Ev 28.

200. Flarup M. Letter to Phil Jones; 4 March 2010. Available at: http://wattsupwiththat.files.wordpress.com/2010/03/data_from_the_hadcrut_dataset_100304.pdf. Archived at: http://www.webcitation.org/632nmlzMN.

201. Stockholm Initiative. Climate scientist delivers false statement in parliament enquiry; 5 March 2010. Available at: http://climateaudit.org/2010/03/05/phil-jones-called-out-by-swedes-on-data-availability/. Archived at: http://www.webcitation.org/632nsHuFP.

202. STCE I; Ev 39.

203. Thorpe A. Email to Sir John Beddington; 6 February 2010. Available at: http://www.whatdotheyknow.com/request/65117/response/175119/attach/7/EIR%20IR110397%20email%20dated%2006022010.pdf. Archived at: http://www.webcitation.org/5zj2HJ92V.

204. Acton E. Email to Sir Muir Russell; 7 February 2010. Available at: http://www.whatdotheyknow.com/request/69023/response/185394/attach/4/Appendix%20A%20Data%20file.zip. Archived at: http://www.webcitation.org/5ziHNfyqz. See file named Russell-Acton&Granatt_100209_red.

205. Davies T. Email to Sir John Beddington; 10 February 2011. Available at: http://www.whatdotheyknow.com/request/33112/response/103013/attach/5/Beddington%20DaviesT%20100211x.pdf. Archived at: http://www.webcitation.org/5ziyZxl1G.

206. Davies T. Email to Sir John Beddington; 4 March 2010. Available at: http://www.whatdotheyknow.com/request/33112/response/103013/attach/6/Beddington%20DaviesT%20100309x.pdf. Archived at: http://www.webcitation.org/5ziz69qQy.

207. Department of Business, Innovation and Skills. Response to author's FOI request; 19 May 2010. Available at: http://www.whatdotheyknow.com/request/lord_oxburghs_report. Archived at: http://www.webcitation.org/5wvkvT8Gg.

208. BHE; pp. 1–2.

209. Oxburgh R. Interview with BBC Radio; 14 April 2010. Available at: http://news.bbc.co.uk/1/hi/8618024.stm.

210. BHE; p. 5.

211. Beddington J. Email to Trevor Davies; 9 March 2010. Available at: http://www.whatdotheyknow.com/request/33112/response/103013/attach/6/Beddington%20DaviesT%20100309x.pdf. Archived at: http://www.webcitation.org/5ziz69qQy.

212. BHE; p. 7.

213. Webster B. Lord Oxburgh, the climate science peer, 'has a conflict of interest'; *Times*; 23 March 2010. Available at: http://www.timesonline.co.uk/tol/news/environment/article7071751.ece. Archived at: http://www.webcitation.org/5zjGDW0BT.

214. Emanuel K. Debate at MIT; 10 December 2009. Available at: http://mitworld.mit.edu/video/730. Archived at: http://www.webcitation.org/5zj5R44MQ.

215. Harvey F. UEA emails – scientific panel announced; *Financial Times Energy Source* blog; 22 March 2010. Available at: http://blogs.ft.com/energy-source/2010/03/22/uea-emails-scientific-panel-announced/. Archived at: http://www.webcitation.org/60FVTlHDj.

216. Cressey D. Another day, another 'climate-gate' inquiry; *Nature* news blog; 22 March 2010. Available at: http://blogs.nature.com/news/2010/03/another_day_another_climategat_1.html. Archived at: http://www.webcitation.org/60FVcIdJl.

217. McIntyre S. Another tainted inquiry; *Climate Audit* blog; 23 March 2010. Available at: http://climateaudit.org/2010/03/23/another-tainted-inquiry/. Archived at: http://www.webcitation.org/5zj5hQnoR.

218. Kelly M. Input to the CRU review; 25 March 2010. Reproduced in DHA p. 81.

219. Beddington J. Email to Trevor Davies; 23 March 2010. Available at: http://www.whatdotheyknow.com/request/33104/response/87056/attach/3/FOI%20100744%20attachment%201.pdf. Archived at: http://www.webcitation.org/60FZcTePe.

220. Science Media Centre. Expert reaction to the Oxburgh report on UEA Climatic Research Unit; 14 April 2010. Available at: http://www.sciencemediacentre.org/pages/press_releases/10-04-14_oxburgh.htm. Archived at: http://www.webcitation.org/60FdLE7tq.

221. Curry J. Interview with Keith Kloor; *Collide-a-scape* blog; 23 April 2010. Available at: http://www.collide-a-scape.com/2010/04/23/an-inconvenient-provocateur/. Archived at: http://www.webcitation.org/5zjA4MUkb.

222. McIntyre S. Oxburgh's trick to hide the Trick; *Climate Audit* blog; 14 April 2010. Available at: http://climateaudit.org/2010/04/14/oxburghs-trick-to-hide-the-trick/. Archived at: http://www.webcitation.org/5zjAHgZA0.

223. Pearce F. Strange case of moving weather posts and a scientist under siege; *Guardian*; 1 February 2010. Available at: http://www.guardian.co.uk/environment/2010/feb/01/dispute-weather-fraud. Archived at: http://www.webcitation.org/61lTVnvYh.

224. STCE I; Ev 31.

225. Chu H. Panel clears researchers in 'Climategate' controversy'. *Los Angeles Times*; 15 April 2010. Available at: http://articles.latimes.com/2010/apr/15/world/la-fg-climate-data15-2010apr15. Archived at: http://www.webcitation.org/681PIQoQl.

226. Gray L. Hockey Stick graph was exaggerated. *Daily Telegraph*; 14 April 2010. Available at: http://www.telegraph.co.uk/earth/environment/climatechange/7589897/Hockey-stick-graph-was-exaggerated.html. Archived at: http://www.webcitation.org/666d0nIfc.

227. Rees M. Statement on the Oxburgh Inquiry; 15 April 2010. The words quoted are the full text of the statement, which was issued to the author by the Royal Society Press Office.

228. Royal Society Press Office. Statement on the Oxburgh Inquiry; 16 April 2010. The words quoted are again the full text of the statement. It is perhaps significant that this second statement was not signed by Rees.

229. Rees M. Letter to the author; 6 May 2010.

230. Oxburgh R. Email to Steve McIntyre; 3 June 2010. Quoted at: http:// climateaudit.org/2010/06/04/oxburgh-refuses-to-answer/. Archived at: http:// www.webcitation.org/5zjGlbDo0.

231. McIntyre S. Oxburgh and the Jones admission. *Climate Audit* blog; 1 July 2010. Available at: http://climateaudit.org/2010/07/01/oxburgh-and-the-jones-admission. Archived at: http://www.webcitation.org/5x1pkcoBn.

232. BBC *Horizon*. Science under attack; Broadcast 24 January 2011. Unofficial transcript at: https://sites.google.com/site/mytranscriptbox/home/20110124_hz. Archived at: http://www.webcitation.org/61vvNZdbT.

233. Climate Change Emails Team. Minutes of meeting with Phil Jones, Tim Osborn and Ian Harris; 4 March 2010. Available at: http://www.cce-review.org/evidence/UEA-CRU_IV5_040310v6.pdf. Archived at: http://www.webcitation.org/626eFVOuv.

234. CG2: 0502; 1 February 2005.

235. CG2: 1897; 8 December 2008.

236. Climate Change Emails Team. Minutes of meeting with David Palmer and Jonathan Colam-French; 30 March 2010. Available at: http: //www.cce-review.org/evidence/Note%20of%20meeting%20w%20David% 20Palmer%20and%20Jonathan%20Colam%20French.pdf. Archived at: http://www.webcitation.org/5xCWvqZiL.

237. Russell M, Eyton D. Notes of Interview with Rob Bell and Laura McGonagle; 26 March 2010. Available at: http://www.cce-review.org/evidence/26%20March%20Bell%20&%20McGonagle%20interview.pdf. Archived at: http: //www.webcitation.org/60FjKPyap.

238. Russell M, Eyton D. Notes of Interview with Ian McCormick and Alan Walker; 26 Marh 2010. Available at: http://www.cce-review.org/evidence/26% 20March%20McCormick%20&%20Walker%20interview.pdf. Archived at: http:// www.webcitation.org/60FjVYkR5.

239. CG1: 0826209667; 6 March 1996.

240. Boulton G. Summary of salient points of interviews with Professors Philip Jones and Keith Briffa, Dr Tim Osborn and Tom Melvin; 9 April 2010. Available at: http://www.cce-review.org/evidence/salient_points_of_9_April_meeting_at_UEA_Final%20(2).pdf. Archived at: http://www.webcitation.org/5xBYCs52r.

241. Climatic Research Unit. Response to 'Summary of salient points...' and additional requested information; 16 June 2010. Available at: http://www.cce-review.org/evidence/Responses_salient_points_April9.pdf. Archived at: http:// www.webcitation.org/5xRyZVPCj.

242. Climatic Research Unit. Response to McKitrick's article in the *Financial Post*; 17 June 2010. Available at: http://www.cce-review.org/evidence/17%20June%20CRU%20comments%20on%20McKitricks%20FT%20article.pdf. Archived at: http://www.webcitation.org/5xTQoFg5m.

243. McKitrick R. Defects in key climate data are uncovered; *National Post* online; 1 October 2009. Available at: http://network.nationalpost.com/np/blogs/fpcomment/archive/2009/10/01/ross-mckitrick-defects-in-key-climate-data-are-uncovered.aspx. Archived at: http://web.archive.org/web/20100121065218/http://www.financialpost.com/opinion/story.html?id=2056988.

244. Climatic Research Unit. Submission by the Climatic Research Unit to the Independent Climate Change Email Review; 1 March 2010. Available at: http://www.cce-review.org/evidence/Climatic_Research_Unit.pdf. Archived at: http://www.webcitation.org/5xFc5irpM.

245. McIntyre S. New light on 'hide the decline'. *Climate Audit* blog; 15 March 2011. Available at: http://climateaudit.org/2011/03/15/new-light-on-hide-the-decline/. Archived at: http://www.webcitation.org/5xFbzaSVg.

246. Houghton J. Climate Change 2001: the Scientific Basis (Appendix III); 2001. Available at: http://www.grida.no/climate/ipcc_tar/wg1/pdf/TAR-APPENDICES.pdf. Archived at: http://www.webcitation.org/5xPOJvfCK.

247. Jones P, New M, Parker D, Martin S, Rigor I. Surface air temperature and its changes over the past 150 years. *Reviews of Geophysics*. 1999;37:173–199.

248. Briffa K, Osborn T. Seeing the Wood from the Trees. Science. 1999;284:926.

249. CG1: 1143819006; 31 March 2006.

250. McIntyre S. Difference in Yamal versions 'not insignificant'. *Climate Audit* blog; 4 January 2010. Available at: http://climateaudit.org/2010/01/04/difference-in-yamal-versions-not-insignificant/. Archived at: http://www.webcitation.org/5xQEJGWWu.

251. CG1: 1146252894; 28 April 2006.

252. McIntyre S. Submission to the Climate Change E-mail Review; 1 March 2010. Available at: http://www.cce-review.org/evidence/StephenMcIntyre.pdf. Archived at: http://www.webcitation.org/67X4TRsvt.

253. Boulton G. Notes of an Interview with Prof. Phil Jones and Prof. Keith Briffa of the University of East Anglia (UEA) Climate Research Unit (CRU); 27 January 2010. Available at: http://www.cce-review.org/evidence/UEA-CRU_IV4_Briffa%20and%20Jones_v6_1.pdf. Archived at: http://www.webcitation.org/67JFF7qZu.

254. Hantemirov RM, Shiyatov SG. A continuous multimillennial ring-width chronology in Yamal, northwestern Siberia. *Holocene*. 2002;12:717.

255. Hegerl GC, Crowley TJ, Allen M, Hyde WT, Pollack HN, Smerdon J, et al. Detection of human influence on a new, validated 1500-year temperature reconstruction. *Journal of Climate*. 2007;20:650–666.

256. RR; p. 133.

257. RR; p. 137.

258. CG1: 1077829152; 26 February 2004.

259. CG1: 1122669035; 29 July 2005.

260. CG1: 1054756929; 4 June 2003.

261. CG1: 1054748574; 28 May 2003.

262. Briffa K. Copies of communications relating to Keith Briffa's editorial treatment of a submitted manuscript; 2 July 2010. Available at: http://www.cce-review. org/evidence/02%20July%20Briffa%20Manuscript.pdf. Archived at: http://www. webcitation.org/63tztf366.

263. CG2: 0286; 24 July 2003.

264. CG2: 1096; 26 September 2003.

265. CG1: 1080742144; 31 March 2004.

266. CG1: 1089318616; 8 July 2004.

267. Boulton G, Jones P. Correspondence related to the Russell Inquiry; April 2010. Available at: http://www.cce-review.org/evidence/15%20April%20Jones% 20follow%20up.pdf. Archived at: http://www.webcitation.org/5xCk4Hbtz.

268. Boulton G. Evidence from Review Editors for Chapters 3 and 6 of the IPCC Fourth Assessment Report on 'The Physical Science Basis'; 1 June 2010. Available at: http://www.cce-review.org/evidence/02%20July%20IPCC% 20fourth%20assessment%20report.pdf. Archived at: http://www.webcitation. org/63vDflQYY. The document does not identify the interviewer, but according to the Russell report, interviews were conducted by Boulton.

269. Intergovernmental Panel on Climate Change. Summary Description of the IPCC Process; 22 December 2009. Available at: https://www.ipcc-wg1.unibe.ch/ statement/WGIsummary22122009.pdf. Archived at: http://www.webcitation. org/63vH3KRtc.

270. Intergovernmental Panel on Climate Change. Procedures for the preparation, review, acceptance, adoption, approval and publication of IPCC reports; September 2008. Available at: http://www.ipcc.ch/pdf/ipcc-principles/ipcc-principles-appendix-a.pdf. Archived at: http://www.webcitation.org/63vJf1vIu.

271. Holland D, Hardie W. Email correspondence relating to Holland's submission to the Russell inquiry; 25 February 2010–7 May 2010.

272. Boulton G, Williams L, Briffa K. Correspondence regarding the Wahl and Ammann affair; 6 May 2010. Available at: http://www.whatdotheyknow.com/request/ what_led_to_boultons_12_may_lett. A zip file of the correspondence is archived at: http://www.webcitation.org/65RoRo4XX.

273. Boulton G. Letter to Briffa and annex; 6 May 2010. Available at: http://www. bishop-hill.net/inquiries-files/20100506BoultonLetterAndAnnex.pdf. Archived at: http://www.webcitation.org/67JI9TopU.

274. CG1: 1189722851; 12 September 2007. Holland quoted this message at paragraph 64 of his submission to Russell.

275. Norton J. Email to Julian Gregory; 4 March 2010. Available at: http://www.whatdotheyknow.com/request/73614/response/197691/attach/ 9/r%20UEA%20CRU%20Investigation%20Mar%2010.pdf. Archived at: http://www.webcitation.org/60aiJSSef.

276. Norton J. Email to Julian Gregory; 22–23 March 2010. Available at: http://www.whatdotheyknow.com/request/73614/response/198085/attach/ 7/r%20UEA%20CRU%20Investigation%20Mar10%202.pdf. Archived at: http:// www.webcitation.org/648xGP0qT.

277. RR;. P. 146.

278. RR; p. 146.

279. Sommer P. Initial Report and commentary on email examination; 17 May 2010. Available at:http://www.cce-review.org/evidence/Report%20on% 20email%20extraction.pdf. Archived at: http://www.webcitation.org/64991rMnj.

280. Palmer D. Letter to David Holland; 26 January 2010. Available at: http://www.whatdotheyknow.com/request/23907/response/67506/ attach/2/Response%20letter%20174%20100126.pdf. Archived at: http: //www.webcitation.org/63vmZqxKS.

281. Palmer D. Letter to David Holland; 26 March 2010. Available at: http://www. ventalize.org.uk/FoI/UEA/Release_letter_Holland_174_100326.pdf. Archived at: http://www.webcitation.org/67Ej6rRoi.

282. Holland D. Personal communication; 14 December 2011.

283. Palmer D. Letter to Andrew Battersby; 13 May 2010. Available at: http://www. ventalize.org.uk/FoI/UEA/20100413_UEAFOI_09-174_ICO_FER0304640.pdf. Archived at: http://www.webcitation.org/67EjhGJF8.

284. Russell inquiry team. Confirmed Note of Actions from CRU Review Group Meeting (Teleconference); 22 April 2010. Available at: http://www.cce-review.org/pdf/ Confirmed%20Note%2022%20April.pdf. Archived at: http://www.webcitation. org/63tmZ2qbH.

285. McIntyre S. Supplemental Submission to Muir Russell; 6 June 2010. Available at: http://climateaudit.org/2010/06/10/supplemental-submission-to-muir-russell/. Archived at: .

286. Bradley R, Hughes M, Mann M, Oppenheimer M, Santer B, Schmidt G, et al.. Email to Muir Russell; 26 May 2010. Available at: http://www.cce-review.org/evidence/Letter_to_Sir_Muir_Russell_Climate%20Scientists_26%20May.pdf. Archived at: http://www.webcitation.org/63to6OMfh.

287. RR; p. 11.

288. RR; p. 44.

289. RR; p. 45.

290. RR; p. 64.

291. RR; p. 76.

292. RR; p. 84.

293. RR; p. 82.

294. University of East Anglia Press Office. University of East Anglia's Response; 2 September 2010. Available at: http://www.uea.ac.uk/mac/comm/media/press/CRUstatements/independentreviews/UEAreviewresponse. Archived at: http://www.webcitation.org/63vsoQRdJ.

295. RR; p. 13.

296. RR; p. 59.

297. Pearce F. Climategate: No whitewash, but CRU scientists are far from squeaky clean; 7 July 2010. Available at: http://www.guardian.co.uk/environment/cif-green/2010/jul/07/climategate-scientists. Archived at: http://www.webcitation.org/63vujIgSz.

298. Editorial. Without candour, we can't trust climate science; *New Scientist*; 14 July 2010. Available at: http://web.archive.org/web/20100719182711/http://www.newscientist.com/article/mg20727692.900-without-candour-we-cant-trust-climate-science.html.

299. Penn State University. RA-10 Final Investigation Report Involving Dr. Michael E, Mann; 1 July 2010. Available at: http://live.psu.edu/pdf/Final_Investigation_Report.pdf. Archived at: http://www.webcitation.org/63yb94Pmt.

300. Crook C. Climategate and the big green lie. *The Atlantic*. 14 July 2010;Available at: http://www.theatlantic.com/politics/archive/2010/07/climategate-and-the-big-green-lie/59709/. Archived at: http://www.webcitation.org/60CODwU9L.

301. US Department of Commerce, Officer of the Inspector General. Detailed Results ofInquiry Responding to May 26, 2010, Request from Senator Inhofe; 18 February 2011. Available at: http://www.oig.doc.gov/OIGPublications/2011.02.18-IG-to-Inhofe.pdf. Archived at: http://www.webcitation.org/67haN1O20.

302. McIntyre S. Wahl transcript excerpt; *Climate Audit* blog; 8 March 2011. Available at: http://climateaudit.org/2011/03/08/wahl-transcript-excerpt/. Archived at: http://www.webcitation.org/640WNgYvp.

303. Kintisch E. Exclusive: climatologist says he deleted e-mails, but not at Mann's behest. *Science Insider*. 9 March 2011;Available at: http://news.sciencemag.org/scienceinsider/2011/03/exclusive-climatologist-says-he-.html. Archived at: http://www.webcitation.org/640XIY4Xe.

304. Romm J. Michael Mann updates the world on the latest climate science; *Grist*; 30 November 2009. Available at: http://www.grist.org/article/michael-mann-updates-the-world-on-the-latest-climate-science. Archived at: http://www.webcitation.org/640YTknia.

305. Cronin M. PSU professor feels fallout of e-mails. *Tribune-Review*; 3 December 2009. Available at: http://www.pittsburghlive.com/x/pittsburghtrib/focus/s_655948.html.

306. McIntyre S. New information on the Penn State inquiry committee. *Climate Audit* blog; 15 November 2011. Available at: http://climateaudit.org/2011/11/15/new-information-at-penn-state/. Archived at: http://www.webcitation.org/67hazDk9u.

307. Allen WE. Notes on the Brune talk on Mann–Climategate in Boulder; 6 October 2011. Available at: http://wattsupwiththat.com/2011/10/06/notes-on-the-brune-talk-on-mann-climategate-in-boulder/. Archived at: http://www.webcitation.org/67hbDggjF.

308. Harrabin R. Harrabin's Notes: Getting the message; BBC News website; 29 May 2010. Available at: http://www.bbc.co.uk/news/10178454. Archived at: http://www.webcitation.org/60DiGlynO.

309. Harrabin R. Third 'Climategate' inquiry to report. BBC news online; 7 July 2010. Available at: http://news.bbc.co.uk/today/hi/today/newsid_8795000/8795643.stm. Archived at: http://www.webcitation.org/60DkYCX5R.

310. House of Commons Science and Technology Committee. Setting the Scene Inquiry: Minutes of questioning of Martin Rees; 27 July 2010. Available at: http://www.publications.parliament.uk/pa/cm201011/cmselect/cmsctech/uc369-ii/uc36901.htm. Archived at: http://www.webcitation.org/60DnwSdXY.

311. House of Commons Science and Technology Committee. Setting the Scene Inquiry: Minutes of questioning of David Willetts; 22 July 2010. Available at: http://www.publications.parliament.uk/pa/cm201011/cmselect/cmsctech/uc369i/uc36901.htm. Archived at: http://www.webcitation.org/60DnY6f4T.

312. Was 'Climategate' the greatest scandal to hit climate science or a mere storm in a teacup? *Guardian* debate; 14 July 2010. Available at: https://sites.google.com/site/mytranscriptbox/home/20100714_gn. Archived at: http://www.webcitation.org/615SUSyQ6.

313. Mosher S, Fuller T. *Climategate: the* CRU*tape letters*. Self-published; 2010.

314. McIntyre S. The CRU data purge continues; *Climate Audit* blog; 31 July 2009. McIntyre's remarks are in a comment on the thread, which is available at: http://climateaudit.org/2009/07/31/the-cru-data-purge-continues/. Archived at: http://www.webcitation.org/65RotG7pW.

315. Davies T. Email to Sir John Beddington; 4 June 2010. Available at: http://www.whatdotheyknow.com/request/46598/response/123604/attach/9/Davies%20Beddington%20040610%20redacted.pdf. Archived at: http://www.webcitation.org/60EvMqc6d.

316. Various. Discussion on *The Hockey Stick Illusion*. Wikipedia; 2010. Available at: http://en.wikipedia.org/wiki/Talk:The_Hockey_Stick_Illusion/Archive_1. Archived at: http://www.webcitation.org/60Ey1AyfZ.

317. Hewitt J. Review of *The Hockey Stick Illusion*; *Chemistry World*; 2 September 2010. Available at: http://www.rsc.org/chemistryworld/Issues/2010/September/Reviews/ClimateChangeScepticism.asp. Archived at: http://www.webcitation.org/60EzGzaFn.

318. Joyner R. Mean-spirited scepticism (review of *The Hockey Stick Illusion*); *Prospect* website; 23 August 2010. Available at: http://www.prospectmagazine.co.uk/2010/08/mean-spirited-scepticism-montford-hockey-stic/. Archived at: http://www.webcitation.org/60EzPoJUy.

319. 'Tamino'. The Montford delusion. *RealClimate* blog; 22 July 2010. Available at: http://www.realclimate.org/index.php/archives/2010/07/the-montford-delusion/. Archived at: urlhttp://www.webcitation.org/616mDNtJC.

320. McIntyre S. Tamino's trick: Mann bites dog; *Climate Audit* blog; 27 July 2010. Available at: http://climateaudit.org/2010/07/27/taminos-trick-mann-bites-bulldog/. Archived at: http://www.webcitation.org/616mWD9F5.

321. Ward RET. Did climate sceptics mislead the public over the significance of the hacked emails? *Guardian*; 19 August 2010. The original post included a stand-first reading 'Andrew Montford who is conducting an investigation into the UEA inquiry has a history of omitting evidence to suit his arguments'. This was apparently added by the *Guardian* editorial team rather than by Ward himself. The revised article is available at: http://www.guardian.co.uk/environment/cif-green/2010/aug/19/climate-sceptics-mislead-public and archived at: http://www.webcitation.org/616umk74h.

322. HSI; p. 186.

323. Russell S. Email to Graham Stringer; 6 July 2010. Available at: http://www. cce-review.org/evidence/MR_response_to_Stringer.pdf. Archived at: http://www. webcitation.org/616v6Z9fJ.

324. House of Commons Science and Technology Committee. MPS hold follow-up session on university climate research; 8 September 2010. Archived at: http: //www.webcitation.org/616vJraYo.

325. STCE II; Ev 1.

326. McIntyre S. More Oxburgh misrepresentations; *Climate Audit* blog; 10 September 2010. Available at: http://climateaudit.org/2010/09/10/more-oxburgh-misrepresentations/. Archived at: http://www.webcitation.org/60Ghchx0z.

327. STCE II; Ev 2.

328. University of East Anglia. University of East Anglia did not change the brief of the Oxburgh Panel; 11 July 2010. Available at: http://www.uea.ac.uk/ mac/comm/media/press/CRUstatements/harrabinstatement. Archived at: http: //www.webcitation.org/65Rp4irxd.

329. STCE II; Ev 4.

330. Williams L. Email to Steve McIntyre; 16 September 2010.

331. STCE II; Ev 5.

332. STCE II; Ev 3.

333. Flood E, (Clerk to the Science and Technology Committee). Email to the author; 5 October 2011.

334. Kelly M. Email to the author; 6 October 2011.

335. Hardie W, Williams L, Russell M. Correspondence relating to minutes of Russell's visit to UEA on 18 January 2010.; 14–16 June 2010. Available at: http:// www.bishop-hill.net/inquiries-files/20100616Russell-Williams_red.pdf. Archived at: http://www.webcitation.org/61m9lKiI3.

336. Montford A. The press conference. *Bishop Hill* blog; 14 September 2010. Available at: http://bishophill.squarespace.com/blog/2010/9/14/the-press-conference.html. Archived at: http://www.webcitation.org/61nFrEbL2.

337. Orlowski A. Oxburgh at work. The *Register*; 10 September 2010. Available at: http://www.theregister.co.uk/2010/09/10/oxburgh_science_select_committee/ page2.html. Archived at: http://www.webcitation.org/61nFQYQdK.

338. Randerson J. 'Climategate' inquiries were 'highly defective', report for sceptic thinktank rules. *Guardian* website; 14 September 2010. Available at: http://www.guardian.co.uk/environment/2010/sep/14/climategate-inquiries-lawson-report. Archived at: http://www.webcitation.org/65RlVX33H.

339. Pearce F. Montford lands some solid blows in review of 'climategate' inquiries. *Guardian*; 14 September 2010. Available at: http://www.guardian.co.uk/environment/cif-green/2010/sep/14/montford-climategate-gwpf-review. Archived at: http://www.webcitation.org/65RpBMCmr.

340. Williams L, Russell M. Correspondence relating to FOI request 08-31; March–October 2010. Available at: http://www.bishop-hill.net/inquiries-files/20101025Williams-Russell_red.pdf. Archived at: http://www.webcitation.org/64kosjoed.

341. STCE II; Ev W1.

342. STCE I; Ev 42.

343. STCE II; Ev 13.

344. STCE II; Ev 14.

345. University of East Anglia Press Office. Phil Jones comments on questions concerning deletion of emails; 26 July 2010. Available at: http://www.uea.ac.uk/mac/comm/media/press/CRUstatements/philjonescomment. Archived at: http://www.webcitation.org/61yxsBrVJ.

346. CG2: 3939; 12 October 2009.

347. CG2: 2203; 2 October 2009.

348. STCE II; Ev 8.

349. STCE II; Ev 9.

350. STCE II; Ev 12.

351. STCE II; Ev 15.

352. STCE II; Ev 16.

353. STCE II; Ev 17.

354. Harrabin R. Speech to Environment Agency annual conference 2010. Quoted at *Bishop Hill* blog; 29 November 2010. Available at: http://www.bishop-hill.net/blog/2010/11/29/harrabin-climategate-inquiries-inadequate.html. Archived at: http://www.webcitation.org/62BpH3fx0.

355. Editorial. Green view: the shadow of Climategate; *The Economist*; 28 November 2010. Available at: http://www.economist.com/blogs/babbage/2010/11/after_climategate_and_copenhagen. Archived at: http://www.webcitation.org/62Bq4IG9v.

356. STCE II; EvW 11.

357. STCE II; EvW 12.

358. STCE II; p. 39.

359. Emanuel K. Email to Steve McIntyre; 5 June 2010.

360. STCE II; p. 16.

361. Guy A. Email to David Holland; 11 July 2012.

362. Butler R, Margolies E, Smith J, Tyszczuk R, (eds). Culture and climate change: recordings. Cambridge: Shed; 2010. Available at: http://oro.open.ac.uk/ 29900/1/culture_and_climate_change.pdf. Archived at: http://www.webcitation. org/64Q1PVrXZ.

363. Wagner W. Taking responsibility on publishing the controversial paper 'On the misdiagnosis of surface temperature feedbacks from variations in Earth's radiant energy balance' by Spencer and Braswell, Remote Sens. 2011, 3 (8), 1603-1613. *Remote Sensing*. 2011;3(9):2002–2004.

364. Mendick R. 'Climategate' professor Phil Jones awarded £13 million in research grants; 5 December 2009. Available at: http://www.telegraph. co.uk/earth/copenhagen-climate-change-confe/6735846/Climategate- professor-Phil-Jones-awarded-13-million-in-research-grants.html. Archived at: http://www.webcitation.org/67hiBlYpj.

INDEX

Because of the complexities of the allegations and the multiple investigations, the index has been hard to construct. For the principal characters, I have not attempted to record every appearance in the text.

ABOUT THE AUTHOR

Andrew Montford studied chemistry at St Andrews University and has been variously a teacher of English in China, an auditor, an accountant and an IT consultant. He now works in scientific publishing. His first book, *The Hockey Stick Illusion*, was published to widespread acclaim in 2010.

He blogs under the pseudonym Bishop Hill (http://bishop-hill.net) when he finds both time and inclination arriving at the same moment and he lives in rural Scotland with his wife and three children.

Made in the USA
Lexington, KY
01 March 2014